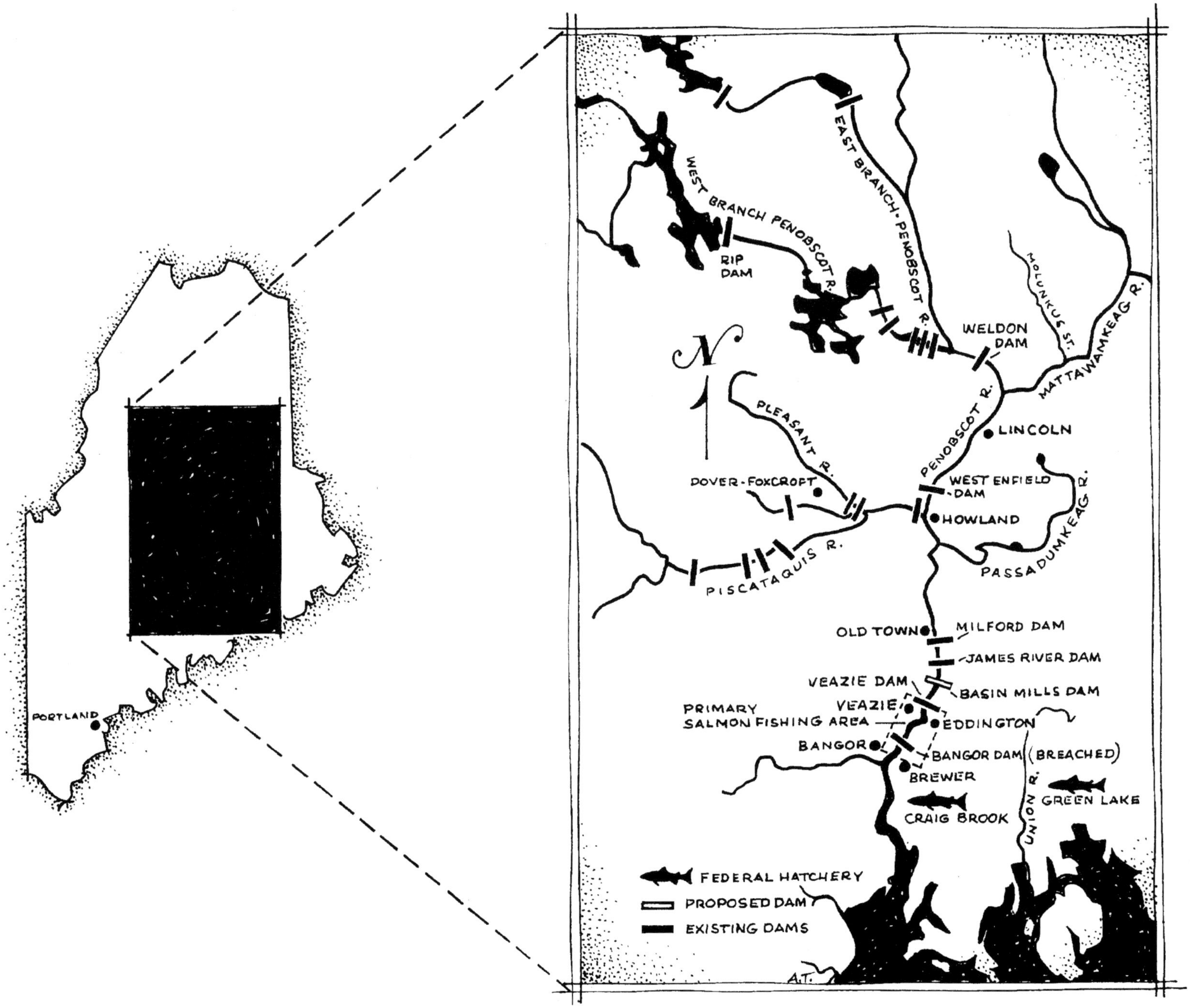

The Penobscot River Watershed

PENOBSCOT RIVER RENAISSANCE

Restoring America's Premier Atlantic Salmon Fishery

This is No. of a special limited edition of 3,000 copies.

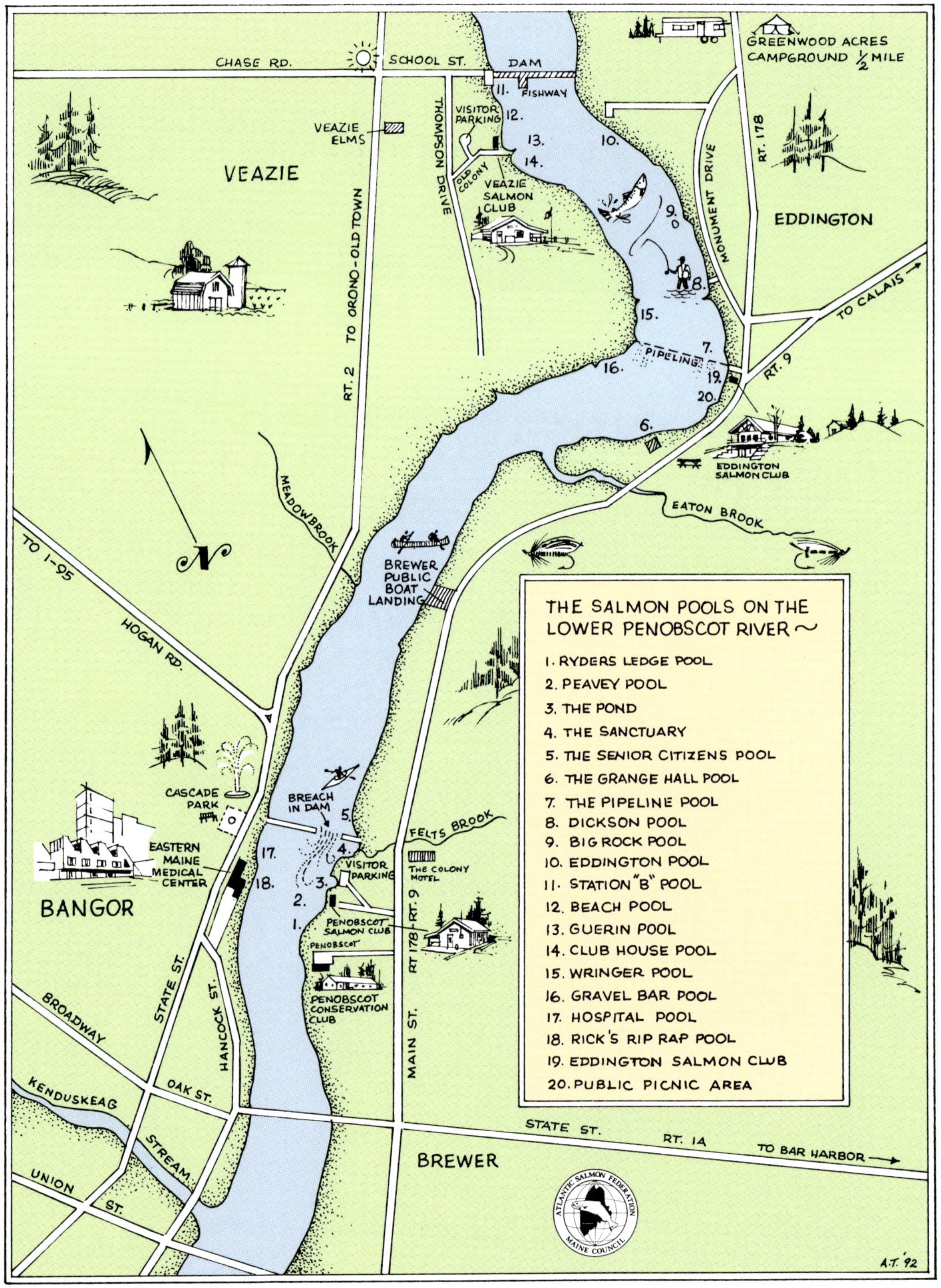
THE SALMON POOLS ON THE LOWER PENOBSCOT RIVER ~
1. RYDERS LEDGE POOL
2. PEAVEY POOL
3. THE POND
4. THE SANCTUARY
5. THE SENIOR CITIZENS POOL
6. THE GRANGE HALL POOL
7. THE PIPELINE POOL
8. DICKSON POOL
9. BIG ROCK POOL
10. EDDINGTON POOL
11. STATION "B" POOL
12. BEACH POOL
13. GUERIN POOL
14. CLUB HOUSE POOL
15. WRINGER POOL
16. GRAVEL BAR POOL
17. HOSPITAL POOL
18. RICK'S RIP RAP POOL
19. EDDINGTON SALMON CLUB
20. PUBLIC PICNIC AREA
CHASE RD.
SCHOOL ST.
DAM
FISHWAY
GREENWOOD ACRES CAMPGROUND 1/2 MILE
VISITOR PARKING
VEAZIE ELMS
VEAZIE
THOMPSON DRIVE
OLD COLONY
VEAZIE SALMON CLUB
MONUMENT DRIVE
RT. 178
EDDINGTON
TO CALAIS
PIPELINE
RT. 9
EDDINGTON SALMON CLUB
EATON BROOK
RT. 2 TO ORONO-OLD TOWN
MEADOWBROOK
N
TO I-95
HOGAN RD.
BREWER PUBLIC BOAT LANDING
CASCADE PARK
BREACH IN DAM
FELTS BROOK
VISITOR PARKING
THE COLONY MOTEL
EASTERN MAINE MEDICAL CENTER
BANGOR
PENOBSCOT SALMON CLUB
PENOBSCOT
RT 178 RT. 9
PENOBSCOT CONSERVATION CLUB
BROADWAY
STATE ST.
HANCOCK ST.
MAIN ST.
KENDUSKEAG
OAK ST.
STREAM
STATE ST.
RT. 1A
TO BAR HARBOR
BREWER
UNION ST.
ATLANTIC SALMON FEDERATION
MAINE COUNCIL
A.T. '92

Penobscot River Renaissance

Restoring America's Premier Atlantic Salmon Fishery

by James E. Butler
and
Arthur Taylor

Paintings by Arthur Taylor

The Silver Quill Press
an imprint of
Down East Books • Camden, Maine

ISBN 0-89272-325-4
Library of Congress Catalog Card Number 92-90314

10 9 8 7 6 5 4 3 2 1

Paintings photographed by Michael Melo
Photographs on page 2, 3, 4, 5 and 6 courtesy Jerry Marancik, US Fish & Wildlife Service
Photographs on pages *v* and *viii* courtesy the Penobscot Salmon Club
Cover and text design by D. Timothy Seymour
Printed in Hong Kong
The Silver Quill Press is an imprint of Down East Books/Camden, Maine

Acknowledgments

Future generations will not treat us kindly if we fail in the task of bringing the Atlantic salmon back to New England. In reality, it's up to us, here and now, to do whatever we can; we may never completely understand the mysteries that surround this magnificent fish, but we can affect what happens to it in our own rivers, right under our very noses. Together, the authors owe our greatest thanks to those who have given their time, money and talent to make sure the Atlantic salmon returns again to the Penobscot River. Without them, there would have been nothing to write about. But we'd like to express our thanks individually as well.

My appreciation goes to the great rivers of Maine and Canada, for all the inspiration they have provided me as an artist. The Penobscot here in Maine, the Margaree in Nova Scotia, the Miramichi in New Brunswick, the Matapedia in Quebec—I hold them all in a special reverence. Also inspiring to me, each in a different way, are the following individuals: Walter Dickson, for his continual involvement, good advice and fine example for us all; Bill Townsend, for his timeless commitment to environmental issues, and his ability to somehow keep us on the right track; Ed Baum, dedicated Atlantic salmon biologist, for defining the priorities for successful river restoration; my good friend and fellow salmon fisherman Bill Taylor, communications director of the Atlantic Salmon Federation, for the awareness he brings to the broad picture of salmon successes and failures; Bob Colson, for his friendship and gentlemanly approach to sportsmanship; Jim Butler, for his enormous contribution of time and effort in writing and pulling all this information together so it made sense; and lastly, I am very grateful to my wife Ruth, whose saintly tolerance for my two major addictions, painting and salmon fishing, is nothing short of miraculous. *—A.T.*

I owe gratitude to many for their encouragement during the making of this book, only a few of whom I can list. My thanks to Allen and Tom Fernald of Down East Enterprise, who had the confidence in me to ask me to take on this project; I hope that confidence has been, at least in a small way, confirmed. Thanks also to Silvio Calabi, editor of *Fly Rod & Reel* Magazine, who understood when I had to disappear from the office to do research. To Nick Lyons of Lyons & Burford, who has encouraged me in innumerable ways throughout my career. To Jerry Marancik of the US Fish & Wildlife Service, who opened his files for our research, and who supplied the bulk of the archive photos in these pages. To Arthur Taylor, without whose guidance I never would have been able to produce my part of this work. His charm and good humor make him wonderful to work with. It's clear why he's befriended by so many anglers of the Penobscot. To my parents, who have always let me be myself; I'm sure that, at times, it's been a tremendous trial. I know I can never repay their love and support. And to Michele, whose determined support and encouragement have made the long hours and occasional, and inevitable, frustrations, a great deal easier. You've never known me without this project dominating my life. I promise that, now that it's finished, you'll discover that I do indeed have a personality. *—J.E.B.*

Dedicated to

Horace P. Bond

We can think of no better words than those
on the granite boulder, inscribed in his memory,
in front of the Penobscot Salmon Club:

He inspired others to work and add their help toward cleaner water, which led to the return of *Salmo salar*, the noble Atlantic salmon.

He did indeed.

Color Plates

Table of Contents

Salmon Flies
of the Penobscot River

Pink Ent
originated by Bob Ent

Early Riser
Ron Newcomb

Verdict
Jerry Clapp

Wringer
Claude Westfall

C. Z. Special
Claude Westfall

Copper Slider
Bill Drexler

Foxfire
Claude Westfall

Eddington Fly
Bob Boucher

Taylor Special
Arthur Taylor

Ghost
Gayland Hachey

Newcomb's Rock
Ron Newcomb

Ray's Red
Ray Stout

Jerry's Green Butt
Jerry Clapp

Turk's Red Butt
Turk Tarini

Ruhlin's Riot
Dick Ruhlin

Dickson Bear
Ron Newcomb

Thunder Head
Ron Newcomb

Mi-Fli
Gayland Hachey

Introduction

Perhaps without knowing it, Charles Atkins started something in the late 19th Century. When a multi-state commission—formed in 1864 to improve depleted inland fisheries—asked Maine fisheries commissioner Atkins to find a site for obtaining Atlantic salmon eggs in Maine, he started by buying live salmon from weir fishermen on the Penobscot River. He stripped eggs from the females, fertilized them with milt from the males and distributed the eggs to other New England states. In a roundabout way, Atkins is thus responsible for the existence of an Atlantic salmon restoration program on the Penobscot itself: The hatchery he established at Craig Brook, near Bucksport, supplies the bulk of the young Atlantic salmon stocked annually in the river.

There is more to the story of salmon restoration than the history of the Craig Brook National Fish Hatchery. But what Atkins created when he bought those first salmon in 1871 was the forerunner of the effort to bring the Penobscot River Atlantic salmon back from virtual extinction.

The story of that effort is one of a persistent search for answers to complex questions. For the Atlantic salmon, often termed the most noble of gamefish, invites complexity. In its life it will swim many miles of river and many, many square miles of ocean, making itself vulnerable to an array of threats. Dams, nets (on the high seas and, in the past, on the Penobscot River itself), pollution, poaching, anglers and natural predators are just a few. With the salmon's propensity to swim international waters and cross international boundaries, it invariably gets wrapped in political wrangling, perhaps more so than any other fish. In the past, its persistence has enabled it to survive. But now we have dammed, poached, netted and angled to such an extent that the Atlantic salmon needs our help.

That's where the Penobscot River restoration comes in. Surely other salmon rivers are in trouble; many runs are either virtually or truly gone, many likely for good. But when the US Fish & Wildlife Service named the Penobscot a model restoration river in the late 1960s, what was done there became critically important to all anadromous fish in the United States, and to all those concerned about them. For if the restoration program on the Penobscot is successful (and we shall see that it really can already be considered so), and the battles against the various threats to the Penobscot's salmon are successful as well, it is hoped it will act as a guide to others fighting similar battles on other threatened rivers and on behalf of other threatened fish.

Charles Atkins' early attempts to supplement the natural run of Atlantic salmon on the Penobscot staved off the threats for a time. But eventually, industrial and residential pollution, nearly unchecked hydropower development, commercial netting and poaching were too much, and the Atlantic salmon was almost completely driven from the Penobscot River (which had historically supported runs of tens of thousands of fish, some approaching 30 pounds in weight).

Not many years after Atkins began his work, in 1882 in fact, a certain Fred Ayer of Bangor showed the world that salmon could be taken on the fly from the river flowing through that city. He caught a salmon in the pool below the Bangor Dam, which was then, and is today, the first dam salmon encountered on their journey up the Penobscot to spawn.

For the next decade, Ayer's name regularly headed the annual lists of top rods at the Bangor Salmon Pool, including the lean years after 1888 when the runs were disappointing (although, apparently, commercial catches on the river had been dwindling since the middle of the century). Local journals attributed the depleted runs to acid and other refuse from the pulp mills upstream, as well as excessive netting on the lower reaches of the river. Stocking apparently helped the situation, although it was noted that the returning fish had blunter snouts and were often smaller than the wild salmon. The work to maintain the Penobscot's Atlantic-salmon sport fishery had begun.

Karl Anderson started something, too, but in 1912. He was the first angler to present a Penobscot salmon to the President of the United States (Anderson's went to William Howard Taft), and he most assuredly knew not what he had wrought. For 42 years thereafter without interruption, the angler lucky enough to catch the first salmon of the season on the Penobscot River sent (or personally presented) his fish to the President. (That needs at least one qualification: On May 2, 1925, Charles Bissell caught a salmon that, instead of sending to President Calvin Coolidge, he sold to Fickett's Fish Market in Bangor. On May 17, Adolph Fischer caught what became the Presidential Salmon that year.) But after Guy Carroll's 10-pounder went to Dwight Eisenhower in 1954, it all stopped. There just weren't any more fish. The Penobscot smelled like a cesspool. A string of industries, from tanning factories to pulp mills, pumped their effluent directly into the flow, to be taken out of sight and thereby out of mind. Towns along the river did the same with their untreated residential waste. A string of dams with ineffective fishways, or no fishways at all, made traveling upriver to spawn a nearly impossible task for anadromous fish of all kinds. High-seas interception of salmon also took a tremendous toll. It wasn't until 1982, when Ivan Mallett presented an eight-pound fish to Ronald Reagan, that the tradition was begun anew.

Part of a great tradition: Horace Chapman presents the first Penobscot salmon of 1931 to President Herbert Hoover.

It is what has happened since Guy Carroll presented that salmon to Eisenhower—the modern restoration effort—that we are concerned with in the bulk of this book (actually, the first faint rumblings had been heard in 1948, with the formation of the Atlantic Sea-Run Salmon Commission). You'll meet men and women who have been, and are, central to the fight against dams, pollution, netting on the high seas and in the estuaries. Of disparate backgrounds, they are eloquent when they speak of their ties to Atlantic salmon and their desire to see it return in numbers to the grand river that flows through Bangor.

They are fishermen (and -women), biologists, politicians, local businesspeople. They often disagree on finer points of the restoration effort, but they all want to see the strong return of the Atlantic salmon. (All but a handful of these interviews took place during a one-year period from early 1991 through the late winter of 1992; most during the 1991 salmon season, with the Penobscot flowing by within sight.) Again, complex questions. Angling regulations, for instance, invariably stir controversy. In late February, 1992, state regulations on salmon were made more restrictive, allowing anglers to kill only one fish per season, either an adult salmon or a grilse. Previously, anglers had been allowed to take as many as five salmon on some rivers. The question for the Atlantic Salmon Commission was whether to risk losing the support of fishermen who want to keep fish, while still performing the necessary task of reducing angling harvest. When the Maine Council of the Atlantic Salmon Federation endorsed a catch-and-release-only policy, it made the one-fish limit a compromise position; you'll hear more than a handful of opinions on the subject here, and there are many more among salmon anglers in general.

Some of the need for stricter regulations may indeed be caused by the restoration effort itself. (That's not to mention the necessity, or political advantages, of such a move. How can anglers ask businesses to give up opportunities—such as building dams—however shortsighted they may be, if the anglers are not willing to make sacrifices as well?) Fish collected as brood stock for the restoration are taken at a trap at the Veazie Dam, just upstream of Bangor; the trap operates best later in the season, when the water is low. Thus, later-returning fish dominate the brood stock, they may produce salmon with a tendency to return late themselves, and so on.

It is said that the salmon run on the Penobscot used to be much earlier than it has been in recent years. Perhaps more important is the effect the fish returning later may have on the restoration of a wild run of salmon, which is the ultimate aim of all involved. Wild fish are becoming a greater percentage of total returns every year (indicating that fry stocking and natural reproduction is increasingly successful), and those fish generally return earlier in the spring—when water temperatures are lower and salmon are more apt to take a fly. Thus, it's more likely that salmon kept are wild salmon; why not reduce harvest of the salmon you most want to have in the river?

Certainly the 1991 return to the Penobscot was late; however, it's easy to over-analyze the minutiae of every year's return, losing sight of the larger, more important issue. As Bill Townsend, president of the Maine Council of the ASF says, "The restoration is a long-term process"; what may be temporary setbacks brought about by natural conditions have to be taken in the context of a long-term commitment to bring a wild run of salmon back to the Penobscot River.

To understand that commitment, we'll look at the history of the restoration (including early efforts), examine what brought the Penobscot to the lows it experienced for two decades beginning in the early 1950s, and learn of the tremendous amount of work that still needs to be done. We'll also enjoy the watercolors and line drawings of Arthur Taylor, one of the Penobscot's premier artists. I hope that, after admiring Arthur's fine art and listening to the anglers, biologists, legislators and conservationists tell their stories of the Penobscot, readers will feel compelled to support the effort to reestablish Atlantic salmon in the rivers of the United States, get involved with the protection of their own home rivers, and find ideas for their own fights against pollution, over-fishing and misguided development.

Jim Butler
Spring, 1992

Opening Day at the Penobscot Salmon Club
1946

The Penobscot, Then and Now

The Native Americans who lived on and near the Penobscot River, as early as 7000 B.C., must have been enthralled with its bounty. But the fish in the Penobscot's waters and the game in the surrounding rolling countryside were not what first brought Europeans to the region. Sebastian Cabot was the first European to see the Penobscot, in 1498, and leave a written record; those who followed were in search of the Indian city of Norumbega, reportedly filled with gold and gems.

They never found it. But little more than a century later—Champlain managed to navigate the river to the head of tide, at what is now Bangor, in 1604—explorers were beginning to realize that they had, indeed, discovered a treasure of tremendous worth. In 1614 Captain John Smith (yes, *that* Captain John Smith) explored the Maine coast and recommended Penobscot Bay to James I as a likely place for a colony. It wasn't the finny occupants of the rivers and streams that were so valuable—it was Maine's forests. During James' reign, the first shipment of Maine pine crossed the Atlantic, to be made into masts for his majesty's navy. The wholesale exploitation of Maine's resources had begun.

By the time it was over, Bangor would be the lumber capital of the world; from the 1830s to the 1880s, about 8.5 billion feet of timber would float down the Penobscot, to be loaded onto ships hailing from ports around the world. Mills would spring up along the river, producing lumber and pulp and paper and dumping the waste into the flow. The population would increase dramatically, and towns would pump their own waste into the river as well, to add to the industrial pollution. Dams were built to produce power, first for the mills and then for electricity for sale to Maine residents and to out-of-state utilities.

Each of these industries, and others, had an effect on the Penobscot. (Those effects were by no means all detrimental. A 1959 report by Richard Cutting, fishery biologist for the Atlantic Sea-Run Salmon Commission, stated, "The extensive lake system with associated dams in the Penobscot drainage has very favorable effects on the river discharge. Effective regulation produces lower maximum flows and increases the summer discharge rates when flows are normally minimal. The producers of hydroelectric power are responsible for greater stability of the stream flows.") But still, logs clogged the streams during the drives (the largest raft recorded contained 113,327 board feet), and sawdust pollution was at its worst in the middle of the last century, when Bangor exported more lumber than any other city in the world. The high point was reached in 1872, when 250 million feet of lumber left Bangor by ship. Sawdust and wood chips from the mills coated the stream bottoms, destroying spawning habitat and killing insects. Perhaps even more important, decomposition of wood waste depleted the water's dissolved oxygen content; by the middle of the twentieth century dissolved oxygen levels would plummet to such a critical point that conditions during the summer months would at some places be entirely inhospitable to fish. By the 1940s and '50s, a zone of zero oxygen reportedly existed below the Bangor Dam during the summer, nearly eliminating the upstream passage of salmon. Deposits of wood chips and sawdust, many feet deep and some a century old, still exist along the Penobscot and its tributaries.

Commercial salmon netters on the Penobscot, probably during the 1890s.

But a generous share, if not most, of the pollution eventually came from the pulp mills that began to appear after about 1880, when the bulk of Maine's tall pine forest had been harvested. Some use had to be made of the smaller trees that were left after most of the virgin timber was gone and the pulp and paper industries were there to take up where the lumber mills had left off. Their waste amounted to something much closer to the industrial pollution with which we're familiar today: the chemicals produced during paper production were the primary culprits in making the water of the Penobscot nearly unliveable for salmon or any other fish species. Add to this short list the various effluents from tanneries, slaughter houses, woolen mills, bottling plants and more, and one can understand what the Penobscot was by the middle of this century—a long, narrow cesspool. Richard Cutting's 1959 report estimates that the Penobscot was receiving a pollution load equivalent to the domestic sewage from 2,500,000 people—at the time, nearly three times the population of Maine, and equal to the combined populations (again, at that time), of Maine, New Hampshire, Vermont and Rhode Island. And the sewage treatment, when any did exist, was a far cry from the standards we have today.

At the same time as these practices were helping to destroy the river, man was exploiting the salmon themselves, both for commerce and for sport. The first installation of a trap weir on the Penobscot was reported in 1815—although commercial fishing had started more than 30 years earlier—and by 1818 the citizens of Bangor were concerned enough about the salmon runs to appoint a commission to regulate the commercial fishery on their river.

"Salmon cars," loaded with live adult fish bought from commercial fishermen, are towed upstream to the Craig Brook hatchery in the 1890s.

But it wasn't enough. The burden of heavy commercial fishing, coupled with increasingly polluted water and a growing number of dams impeding upstream and downstream movement of fish, caused salmon returns to drop dramatically. Charles Atkins, first director of the Craig Brook National Fish Hatchery at Bucksport, reported that the commercial catch in 1873 amounted to nearly 15,000 salmon; yet the average annual catch from 1895 to 1905 was fewer than 5,000, and by 1920 it had dwindled to 1,500 fish. In 1947, the last year of legal commercial salmon fishing on the Penobscot River, the take was 40 fish.

Sport fishermen also noticed a deterioration in the salmon runs. By 1889, only seven years after Fred Ayer took the first salmon on a fly from the Bangor Pool, local journals noted that the quality of the fishing was dropping off, and blamed the trouble on "the acid and refuse from the large pulp mills on the river." Stocking efforts brought a resurgence in just a few short years, so that by 1892 anglers took 100 salmon from the pool, approximately the number averaged annually on rod and reel from that year through 1930. (As low as those numbers may sound, we must remember that the sport at the Bangor Pool was in its relative infancy at the time, with only 40 fish caught in 1885, when conditions—and the run—were quite good.)

There were still some good years left—248 fish were landed in the Bangor Pool in 1931—but deterioration came swiftly. The '30s saw an average of about 85 salmon taken each year, but by the '40s the annual average had plummeted to below 20. Only 63 fish—total—were taken from 1951 through '59, including three years of no catch at all. Sport catches on the river as a whole dropped just as precipitously. Clearly, something had to be done.

A technician at the Craig Brook hatchery sorts fertilized eggs from unfertilized (again, probably in the 1890s).

There had been active efforts to bolster the salmon runs for many years, starting in 1871 with Charles Atkins' work at the Craig Brook National Fish Hatchery near Bucksport. Atkins was a true scientist, tagging stocked salmon to gauge returns. (There is probably a master's thesis or two hidden in the voluminous and detailed journals he kept during his more than 40 years in fish culture and now stored at the National Archives and the Craig Brook hatchery.) Under his direction, the Craig Brook hatchery propagated salmon eggs that were shipped as far as Wisconsin and Iowa. Atkins began by buying eggs from salmon culturists in Canada, but the cost was at times as high as $40 per thousand—at such a rate, the eggs from a single adult would cost a few hundred dollars. To cut expenses, he bought salmon from weir fishermen on the Penobscot and kept the fish at Craig Brook until they were ready to spawn, transporting them upriver to the hatchery in wooden scows, known as salmon cars, with holes punched in the sides to keep the water well oxygenated. Atkins then stripped them of eggs and milt and fertilized the eggs using his own derivation of the "Russian dry" method, which essentially involved mixing the eggs and milt in a pan and then adding water. The resulting fry, parr and smolts were stocked in the Penobscot, and became the source of a "new" Penobscot strain when the salmon runs in the river were reduced to almost nil.

Atkins' efforts were admirable, and effective in bringing salmon back to the Penobscot River annually as long as the major obstacles were commercial fishing and dams with something resembling fishways. There would always be a few salmon that could avoid the fishermen's nets and scale the fish ladders to the spawning grounds. And the fish returning to the river could be caught at the Bangor Pool (before their upstream progress was stopped by dams, the downstream-most of which was the Bangor Dam), if nowhere else. But pollution brought the Penobscot to such a marginal level that stocking on even the most enormous scale would not have been enough—salmon will not return to a river that cannot support fish life, no matter how many are stocked and manage to survive their allotted seasons on the high seas.

Opening Day at the Bangor Salmon Pool, likely before the turn of the century.

The members of the Atlantic Salmon Commission realized this when the commission was formed in 1948; yet they had to wait for national and state legislation to make the restoration possible and make their efforts effective. As Dick Ruhlin notes in his interview that follows, only a few years after the Clean Water Act of 1968 was passed and its requirements went into effect, salmon were seen again in the Penobscot. When domestic and industrial sewage was limited, when treatment was required, nature again showed how efficient she is at repairing the damage wrought by man.

Now the fight becomes one to minimize obstruction on the Penobscot and continue to improve water quality. To most, the true aim of the restoration and a critical rationale behind the creation of the Atlantic Salmon Commission is the recreation of a natural, self-sustaining run of Atlantic salmon on the river. The dams now in place—with their inadequate upstream and downstream fish passage—and the new dams and modifications of old ones proposed, pose the biggest obstacle to realizing that goal.

There have been tremendous accomplishments, an example being the defeat of the Swift River Company's proposal to rebuild Bangor Dam after it was allowed to deteriorate to such a point that a breach appeared through the structure itself, allowing salmon to pass upstream. The breach has since increased in size, making fish passage even easier.

The most crucial fight at this writing is Bangor Hydro-Electric Company's proposal to build Basin Mills Dam, near Orono, on the Penobscot. Computer models say that if that dam is built, it will reduce the possibility of reestablishing a legitimate run of salmon to somewhere below 40 percent. Given that, it's possible the federal government will seriously reduce its support of a program that has come remarkably far since the 1950s, when salmon returns were virtually nonexistent.

Spearing an Atlantic salmon during the height of the run—a practice that, while no longer legal, still takes place.

Jerry Marancik, anadromous fish coordinator for the US Department of Fish & Wildlife, thinks the numbers from that computer model have been misused. He contends that the original data is good, but that the variables on which the results are based can easily change over time. It might even improve. But Jerry points out that if there are mistakes in the data, they're consistent in regard to the impact of Basin Mills. For instance, if the model says the building of Basin Mills will reduce the chances of restoring salmon to the Penobscot from about 70 percent to less than 40—as it does now—or from 90 to 60 percent, it still makes clear the detrimental impact of the dam.

Still, the Federal Energy Regulatory Commission (the agency responsible for approval of hydro projects, but in effect a rubber stamp for dams across the country), seems bent on allowing nearly unbridled development in the interests of power generation.

As the various battles have progressed, and the future of the Atlantic salmon on the Penobscot looks brighter, attendant things have occurred: the Penobscot Salmon Club, after suspending operations for a number of years, is active again; two new clubs, the Veazie Salmon Club and the Eddington Salmon Club, have been formed; alliances with national groups not directly connected to salmon have developed, most notably with the Sierra Club Legal Defense Fund, which is providing valuable (and free) legal assistance to the Penobscot Coalition in the fight against Basin Mills; American Rivers has named the Penobscot one of the 10 most threatened rivers in the nation. But we're jumping ahead to the modern Atlantic salmon restoration . . . we're going to let those involved tell the story on the following pages. —*J.E.B.*

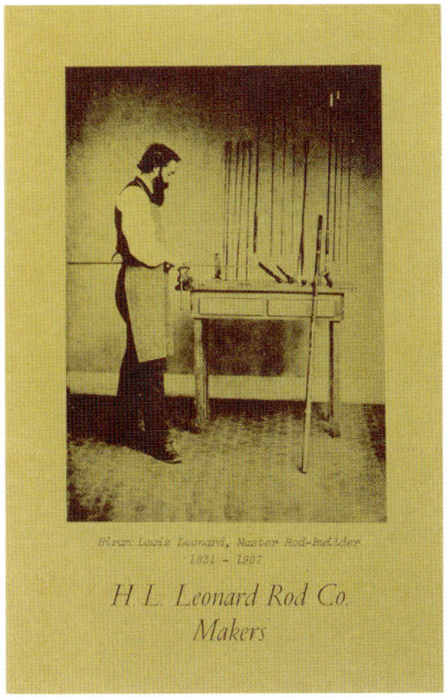

Bangor's Bamboo Legacy

While Atlantic salmon fishing and the more-than-century-long efforts at salmon protection and restoration on the Penobscot are well-documented, relatively few realize the critical role the city on its banks—Bangor—played in the development of modern fly tackle. It was here, in the 1880s, that legendary craftsman Hiram Lewis Leonard—considered by many to be the "father of the fly rod"—began building his innovative six-sided bamboo rods. A handful of other important rod makers got their start under Leonard's tutelage and, by extension, he can even be considered an influence on many modern makers of graphite rods.

An unusual man of many talents, H.L. Leonard was musical by nature; a local hunter and naturalist friend, Manly Hardy, says Leonard always carried a flute with him and often played well into the night on hunting and camping trips. But Leonard was also an expert fisherman and gunsmith, a sometime professional hunter and an engineer by profession. He fashioned his first fishing rod, from lancewood and ash, in 1870 or '71.

Born in Sebec, Maine, in 1831, Leonard's early years were spent helping in his family's oar-making business—he became an engineer after completing his formal education at 15. In 1869, he started a gunsmith business on Main Street in Bangor. Another gunsmith in the area, Francis Philbrook—a very skilled machinist—teamed up with rod maker Ed Payne to produce one of the finest reels in history, the Philbrook & Payne. Philbrook later sold the reel patents to Leonard and continued to manufacture high-quality reels for Leonard, under the Leonard name.

In 1871, H.L. Leonard relocated his small shop on Main Street to a larger one at 2½ Kenduskeag Bridge (near the present-day—and now vacant—Sears building). By 1873 his rod-making was doing so well he phased out the gun business. It was during this period that Leonard invented his famous beveler, a machine that would precision cut bamboo strips of constant uniformity. A revolution in its time, it eliminated hand-planing, and by changing the taper patterns, Leonard could produce any type of rod.

After Leonard moved his operations to Central Valley, New York in 1881 (to be nearer William Mills of retailer William Mills & Son, a partner in the Leonard firm), he hired Bangor craftsmen Fred Thomas and Eustis Edwards. They joined Leonard in Central Valley, and Ed Payne, also from Bangor, came on board in 1885. Tensions caused by the speed of work necessary to fill orders (they had never been greater) convinced Thomas, Edwards and Payne to leave in 1889 or 1890 and form the Kosmic Rod Company. The partnership was short-lived, and in 1898 both Thomas and Edwards returned to Brewer, Maine, where they formed Thomas & Edwards. This combination, too, lasted only a short time, and when Edwards left, the F.E. Thomas Rod Company was born in Brewer, to move by 1902 to the Stetson Building at 119 Exchange Street, Bangor.

When Eustis Edwards took his new company, Edwards & Sons, to Connecticut, he left the F.E. Thomas Rod Company alone in Bangor, where it prospered. The various grades of Thomas rods (the Bangor and Dirigo models, the Browntone and the famous Thomas Special) gained wide acceptance in the sporting world.

In 1923, the company moved into a new brick building at Dark and Center streets in Bangor, a building that is now the Richard Parks gallery. Fred's son, Leon, was serving his apprenticeship under his father's watchful eye, and would eventually take the reins of the company.

Leon employed only the most highly skilled people he could find in and around Bangor. One of them, Ed Houston of Brewer, would go on to become the shop foreman, and he carried on the Thomas tradition for many years after the demise of the company in the middle 1950s. With World War II over, the fiberglass rod rapidly replaced the more expensive and labor-intensive cane rod—and the Thomas Rod Company's beveler would soon to turn out its last strip for a Thomas rod.

In 1958 Leon Thomas sold the company to Sam Carlson of Essex, Connecticut, a superb craftsman and former employee of Gene Edwards of quad-rod fame. The beveler went to Walt Carpenter, a young and very talented rod builder now living in Pennsylvania, and it is once again turning out strips, now for fine Carpenter rods.

In the 1970s, Sam Carlson revived the Thomas name for a short time, making what he called "Thomas Fours" because they were four-sided, unlike the six-sided rods produced by the F.E. Thomas company.

Many Thomas rods built during the 60-year history of the F.E. Thomas company are still on the water today, handed down from generation to generation after years of use. They are prized collector's items, along with Leonards, Edwards and Philbrook & Payne reels. Interestingly, all got their start in Bangor, almost within sight of the Penobscot River.

Today what remains of the F.E. Thomas Rod Company—tools, rod blanks—rests in the Maine State Museum in Augusta, donated by Sam Carlson. He is retired and living in New Hampshire; Hiram Leonard played his flute in 1892 and Carlson enjoys singing with a Swedish vocal group in 1992. Could it be that music is integral to the rod maker's art—or is it the reverse? —*A.T.*

Pipeline Pool at Eddington

From the collection of Jim Taylor

The Interviews

Bill Townsend

President of the Skowhegan, Maine, law firm of Perkins, Townsend, Shay & Brown, Bill is president of the Maine Council of ASF, and one of three US commissioners in the North Atlantic Salmon Conservation Organization. One of Maine's most active (and most honored) environmentalists, he is past president of the Natural Resources Council of Maine and a past member of a half dozen advisory groups to the state government. He is the recipient of, among others, the Sol Feinstone Environmental Award, the ASF Lee Wulff Salmon Conservation Award and, in 1992, Down East *Magazine's Conservation Award.*

I got started salmon fishing in the late '70s. Bucky Owen, who is head of the wildlife department at the University of Maine, and I had known each other for pushing 25 years, and he told me I ought to come over and try for salmon. I did, and was lucky enough to hook a couple of fish early on. When the Maine Council was formed by the Atlantic Salmon Federation in 1982, he asked me if I'd come to the original organizational meeting, which was held right here in Bangor. I did, and I became a delegate through the Friends of the Penobscot River, the group that was fighting the rebuilding of the Bangor Dam at the time. I was a delegate for a couple of years, became a member of the board of directors for a couple of years, then vice president, and then president, which I've been for a little over three years.

I don't fish as much as I'd like anymore. I haven't fished the Downeast rivers at all in the past few years. It's an awful thing when you think, "Gee, I really ought not to go fishing today because I ought to sit down and write this letter . . . ." But you know, somebody's got to do it or it doesn't get done.

I got involved with the NASCO business [North Atlantic Salmon Conservation Organization] because one of the original commissioners, Dick Buck, resigned a year ago last April, and there never had been a commissioner from Maine. Alan Peterson, who had been the regional director of the National Marine Fisheries Service, used to come down salmon fishing in Washington County—he was a member of the Dennys River Club—and had some Maine connections. Dick Buck is from New Hampshire, and Frank Carlton is from Savannah, Georgia, which seems like a very unlikely place for a salmon fisherman or commissioner to come from. But he's been involved with marine fisheries for a great many years. He was involved with the tuna commission, and he has done some salmon fishing.

They were the three original US commissioners to NASCO, and when Dick Buck resigned there was quite a lot of competition for it. I had a very simple platform. My platform was "Maine is where the salmon are. Maine needs a NASCO commissioner." Everybody said, "Yeah, that sounds right," and fortunately I had some fairly good political connections with both [senators] George Mitchell and Bill Cohen. They were an enormous help. I've been involved in the environment since 1960, and I got a lot of my environmental friends to put in a good word. This is a very mysterious process, because it's a personal decision by the President, and it's obviously very, very political. I got good support from the agencies—the Fish & Wildlife Service and our own people here in Maine with the Department of Inland Fisheries & Wildlife, the Department of Marine Resources and the Atlantic Sea-Run Salmon Commission, as well as the National Marine Fisheries Service—so here I am, a NASCO commissioner.

I'll give you an example of how the politics work. This year the Greenlanders proposed a quota that we thought was too high. The Canadians proposed a quota that we also felt was too high. We opposed both. The European Economic Community, which represents the United Kingdom, Ireland, Spain and France, which are the four salmon-producing countries of Europe besides the Scandinavian countries, abstained. On the surface that doesn't look like much, but the fact of the matter is that the Greenlanders are really upset at the lack of scientific information that's coming from the EEC. That's what's behind a lot of this. The reason is that there are some very powerful political forces in Europe that are still making money off the salmon and that want their commercial fishery. They're squashing the hard data. Eventually we have to unravel all those kinds of things and find what buttons to push. The politics are as arcane as anything I've ever seen. The only thing you can do is figure out what's best for the resource and let the chips fall where they may.

As I've been fond of saying lately, in NASCO there's no judge, there's no referee saying that there are certain rules that we have to follow. These international negotiations are essentially every nation looking out for its own interests, but in the context, one hopes anyway, of all of them realizing it's in their self-interest that we protect this resource. But everybody has a different perception of what is in their self-interest.

Consequently, we didn't get a quota agreement on commercial harvest this year. The Greenlanders were willing to come down to 830 tons from 840 tons; we thought that was ridiculous. The Canadians were willing to have about 720 tons, and we thought *that* was too much. We felt that neither the Greenland nor Canadian proposal was satisfactory, so we voted against both of them and came away without a quota agreement. I hope that people will come to their senses in another year and agree. Obviously the Greenlanders can't keep creaming a resource and expect the resource to be there for them to continue. And they know that.

As a matter of fact, it's pretty obvious to me that the commercial interests in the European countries are represented by their commissioners, although they don't say so. We're the only country around the Atlantic that has no commercial fishery. The Canadians have a commercial fishery, and they're looking out for their commercial as well as recreational interests, although next year we think they won't have a commercial fishery, except in Labrador. They'll be completely recreational, which will mean a big buyout of licenses. That should take a lot of pressure off our fish. I would say we're getting four out of every 10 adult fish back, and of every six fish taken in the high seas commercial fishery, probably four are taken by the Greenlanders and two by the Canadians, although it varies from year to year.

Things on the Penobscot are going more slowly than I would like, I think more slowly than anyone who's involved in it would like. As you know, the restoration is a long-term process for the Fish & Wildlife Service and the Atlantic Salmon Commission; they're looking 30 years in the future. They know it isn't going to happen overnight. The high seas commercial fishery is clearly one of the problems. Predation on smolts in the estuaries is another problem. The problem of downstream passage through the dams for wild smolts—and after all, what the restoration is trying to do is establish wild fish again—is a big one, and I think the utilities are really dragging their feet. Upstream passage is another problem—right here at Veazie Dam, for instance, where Bangor Hydro has stalled for at least five years on putting in new upstream fish passage unless they hook it in with the construction of Basin Mills Dam. There are a lot of problems. Yet, if you look at the figures, the long-term trend is up and the percentages of wild fish, long-term, are up. Like everything else, this is no business for the short of breath or the faint of heart. You've just got to keep plugging.

They talk about "state-of-the-art" downstream passage, but state-of-the-art just happens to be as good as you know how to do it right now. They are working on it, and of course the downstream-migrating smolts go down through the top of the water column, which is good, because it makes getting them over the dam easier. But all the water that goes anywhere other than through the turbines is, from the power company's point of view, wasted. They want to have as little spill as possible, and we want to have as much as is needed for getting the fish down.

We're trying all kinds of techniques to encourage the smolts to go to the spillways, whether they're pipes or over the top of the dam, as opposed to going through the turbines. Turbines aren't necessarily fatal, but the percentages aren't good. Sometimes they'll chew up a whole run. There have been times here on the Penobscot when they've felt that every fish coming down was turned into gull bait. The power companies just have to spill water. They think they have the rivers to themselves, and they don't.

We've had a lot of discussions on Basin Mills as to whether we should try to split the Veazie application [for C-Station, an additional power station] away from the Basin Mills application, and during our last discussion of it, which was maybe two or three months ago, we decided that would be a diversion of our effort. We've only got limited resources to deal with a host of problems. Just the same, I think it's an outrage that Veazie didn't have good upstream fish passage at the B-Station five years ago. It's just been deferred and deferred and deferred because of Basin Mills. At the same time, mealymouthed Bangor Hydro is saying, "Restoration isn't going to work." But part of the reason it isn't working now as well as it should be is because they don't have decent upstream and downstream passage.

This river has already given to the maximum, with all the hydro on it. There should not be any more dams on the Penobscot. What the power companies are hoping is that eventually the restoration agencies in the public sector and the private sector will give up and go away and they'll have the river to themselves.

One of the problems in recent years is that we've had poor water conditions, when the smolts are coming down and when the mature fish are coming back. Whether that's just a cycle that we'll go through and end up having some wetter years, who's to say? But you've got to look at these things in the long term, in five-year blocks, 10-year blocks, 15- and 20-year blocks. I think you just have to take the long view. I think another major issue is forestry practices. Because of the clearcutting and the general impact on the forest, you have a situation where you have a lot of high water over a short period of time, and then you get long stretches of low water. There's not a long, slow release of cold water into this river, which really would be much better for the fish.

To my mind, probably the most encouraging thing about the whole restoration is the change in the attitude of the fishermen. I'm so encouraged by the number of young people who really feel very strongly about turning things around. I have certainly seen, and I think anyone who's been around much has seen, that the fishermen really want it to succeed. They understand what the process is, they understand it's not just the federal government creating a put-and-take fishery so they can catch a fish, it's the restoration of a wild species. I talk to people now who, even two or three years ago, thought that the fish were here for the fishermen to catch, and now they understand that the fish are here to get up on the spawning beds. Maybe eventually if we get a truly wild run of fish there will be an opportunity for fishermen.

The original thinking was there had to be a political base, which is why the fishing here isn't entirely catch-and-release. When this whole thing got started there weren't many salmon fishermen, because there weren't any salmon. I haven't looked at the statistics for licenses, but certainly there has been a tremendous increase. That thinking has hung on, and I think there are certainly many people who think they ought to be able to fish, and they ought to be able to kill a fish. My own view is that there ought to be no fish killed, and I suspect that may be going further than a lot of people are willing to go. I do think at this point that the Salmon Commission is looking very hard at the idea of perhaps no fish killed in May or June. At the same time I think there are a lot of people who could kill a fish and who won't. Like Don Shields, who caught a fish over here this year and released it and said, "Well, I deserve to have a salmon," so he went to the store and bought an aquaculture salmon. That's the right kind of thinking.

I think there's a real need for the private sector to get involved in routine patrols to deter poaching. Just getting out there and being seen on the river. One of the major problems I perceive is so many roads in the woodlands. If it were difficult for the poachers to get to someplace—if they had a six-mile hike to get back into some spawning redd—you can be sure the pressure would be a lot less than it is. But they seem to know all the roads, and how to get around the gates.

The Salmon Commission biologists are dedicated, but they're just like everyone else—they're only allowed to work so many hours in a week. When they were originally set up as a staff, they didn't have all this hydro stuff. Now more than 50 percent of their time is spent dealing with hydro issues because of all the relicensings coming up, and when they're dealing with those issues how can they get out and do anything else?

Study after study over the years show that people are willing to pay to protect their resources. They're willing to pay higher electric rates if that's what it comes to, to pay to have a world that isn't just totally sterile, made of concrete and with no wild things in it. People want wild things—that's why we have eagle restoration, that's why we have caribou restoration, that's why we've got Atlantic salmon restoration.

I think there's some merit to the idea of taxing the hydro companies' use of water. The same cubic foot of water gets used maybe 10 times. The Edwards Dam on the Kennebec River is an example. As a rate-payer, I'm subsidizing the people who own that dam, because they're selling the power to Central Maine Power at a much higher price than Central Maine Power is selling it on the market. If CMP is losing on that, you better believe they're making it up elsewhere.

I'm going to guess most of the grilse in the Penobscot are hatchery fish. The last figures I saw showed a pretty high proportion of grilse, which I don't think is considered particularly desirable. What the reason for the grilsification is—whether it's environmental because of some condition in the sea, or whether it's genetic, or a combination of those—I don't know. But I believe that most of the grilse that we're getting are hatchery fish.

Traditionally there weren't many grilse in the Washington County rivers, and traditionally I don't think there were many grilse in this river. I've heard it suggested that perhaps this run should be managed for grilse. I think that was an idea that didn't get very much support, for pretty obvious reasons. Grilse are nice fish, but they're not salmon.

I'm an optimist, I've always been an optimist. You have to be an optimist to be in this business. You have to think positively. I think back how things were 30 years ago when I got started in the environmental business. Our rivers were sewers. The Penobscot River was an open sewer, the Kennebec River was an open sewer, and everybody just kind of accepted that. That was the way it was, and "Gee, it's too bad, but gosh, that's the way it is and that's the way it's going to be." There was just a small group of people who kept saying, "We've got to clean our rivers up." I remember back in 1963 or '64, I put together a meeting down in Portland to talk about clean water. We called it "What Price Clean Water?" and we had Ed Muskie come as our principal speaker. We had people talking about salmon and other kinds of fish. There was a lot of latent support, but there wasn't much vocal support. Now look. We've got a river that's been cleaned up and that can support Atlantic salmon, and that can support a lot of other fish: alewives, shad, striped bass, sturgeon. That's taken 30 years to accomplish. We're in the same situation with the air. It's taken 20 years to get our air quality headed in the right direction instead of dumping sewage into the sky. If it takes 30 or 50 years to accomplish this goal, that's not too long. It took a lot longer than that to create the mess in the first place.

For Bangor Hydro to say, "Salmon restoration is a failure, so let's cut it off and let us put all the water in the river through the wheels," I think is hypocritical in the extreme, because they're part of the problem. I always say that whenever they talk about the restoration not working, they're like the person who killed his parents and then pleaded to the court for mercy because he was an orphan. I don't have much use for them, I'll be very blunt. If we didn't have the high seas commercial fishery, if we didn't have Bangor Hydro stalling, stalling, stalling, there's no question in my mind but restoration would be light-years ahead of where it is today. Those are the two big problems, and we're gaining on them. There's no question in my mind this Basin Mills Dam is not going to be built. In terms of the commercial fishery, the Greenlanders want to be able to share in the growth of the resource, and that's perfectly acceptable. The whole of NASCO is premised on their being entitled to some sort of so-called grazing fee. But they shouldn't have a 60-percent grazing fee, they should have maybe a 10- or 15-percent grazing fee. If we can get the stocks back where they were, there will be plenty to go around. I'm an optimist, but it's going to take a lot of work, and it'll be going on after I'm dead and buried, I'm sure of it.

Don Shields

Born in Leominster, Massachusetts, Don graduated from the University of Maine in 1956 with a BA in History and Government. Now living in Bangor, he and his wife, Diane, have raised four children. Don is a member of the board of directors of the Maine Council of the Atlantic Salmon Federation, the chairman of the Penobscot River Coalition, and was named Trout Unlimited's New England regional rivers coordinator in early 1992.

I first started salmon fishing in the late '70s. I lived here in the '50s, when I went to the University of Maine. My family's had a camp in Ellsworth since 1963, so I'd come up every summer from New York and then New Hampshire, and every summer in the late '70s I'd go down and try the Bangor Salmon Pool. I didn't really get serious about it until 1984, when I moved up here from New Hampshire and started fishing for salmon on a regular basis. The Wringer Pool, on the Veazie side, is my favorite pool, and it's only 10 minutes from my house. I spend a lot of time there.

I'm known for wading without waders—I just like to get in there with the fish. I call it the Penobscot River high. There was a study done on fishermen who fish in trout streams in the mountains. The cold water passes over their legs, and when the cooled blood from their legs circulates back to their brain, their brain releases endorphins, which give them a high. I get as much of a high out there as I do drinking a couple of brews.

We had a few hundred fish coming back each year in the late '70s; there were about 20 years when there were no salmon caught. I think the last one that was given to the President was to Dwight

Eisenhower in 1954. From then until about '77 there were hardly any fish coming back at all. The Anadromous Fish Act was passed in 1965, and the real drive by the US Fish & Wildlife Service to restore the river, when they started working hand in hand with the Atlantic Sea-Run Salmon Commission (the state agency), was 1967. That's when they built fishways on the river. The big things were to clean up the rivers and put in fishways. There was no fishway at the time at the head-of-the-tide dam, which was the Bangor Dam then. There were no other fishways, as far as I know. There was no Veazie Dam fishway. The Veazie Dam fishway, which is now the head-of-the-tide dam, was built in 1968.

When I moved up here in 1984 I joined Trout Unlimited and put my name in at the salmon clubs. Bangor Hydro put in an application to the Federal Energy Regulatory Commission for the Basin Mills project in 1984, and withdrew it. They reapplied in 1987, and that's when I became really concerned. When they made the first application, the big battle was ending on the Big A Dam on the West Branch of the Penobscot, and the battle was in full swing on the Bangor Dam. The Swift River Company wanted to rebuild it. Friends of the River was formed at that time—that was the coalition that fought Swift River and won the Bangor Dam fight through the state DEP. Those things might have contributed to Bangor Hydro's pulling out of the project, because they could see all the local opposition.

When Bangor Hydro resubmitted the application in 1987, I realized the impact of one more dam on the river would make a big difference to the Atlantic salmon restoration program. In 1986 5,000 salmon came back to the river—the restoration was showing promise. We'd been getting 2,500 to 3,000 a year prior to that. We could see the possibility of 10,000 or 12,000 Atlantic salmon coming back in the future because, as I understand it from the resource agencies, there is that much spawning habitat. The river will support a run of at least 10,000 salmon.

That would be under ideal conditions, and certain things would have to happen. When all these dams come up for renewal in the next three or four years, they're going to have to have best-possible upstream and downstream passage. Of the five dams on the main stem of the river, the only one that has the best-possible downstream passage is the West Enfield Dam. That is only because US Fish & Wildlife insisted on it when Bangor Hydro rebuilt the dam and got it relicensed. Relicensing is usually for 50 years, so this is a once-in-a-lifetime chance for the environmental groups and the resource agencies of the state to get the best-possible fish passage and recreational access.

The Veazie Dam was supposed to be relicensed in 1985, but Bangor Hydro has been dragging their feet and getting a provisional license annually. The Federal Energy Regulatory Commission annually and automatically relicenses the Veazie Dam under the old license, which means Bangor Hydro doesn't have to do anything. It's the responsibility of FERC to make sure this doesn't go on for a long time. FERC is certainly in favor of hydropower, but they don't like to be jerked around too much by hydropower utilities. For example, they were upset when Bangor Hydro, knowing the license for the Veazie Dam was going to come up for renewal in '85, put in an application for the Basin Mills Dam in '84 that included relicensing the Veazie Dam and putting in fish passage. When Bangor Hydro withdrew that application, waited another three years and resubmitted it, FERC wasn't very happy because they thought they were being manipulated.

To build Basin Mills, Bangor Hydro also has to get a permit from the Public Utilities Commission, a state agency. The PUC first denied the permit without prejudice, which means Bangor Hydro could come back in a couple of years. In 1988 the PUC had, under their regulations, fined Bangor Hydro for not instituting conservation programs. When Bangor Hydro came back in 1990 with an application for Basin Mills Dam, they still didn't have any conservation program. So the PUC nailed them. They had an investigation, as a matter of fact, into their management practices in the areas of conservation and energy-efficiency measures. They found Bangor Hydro woefully lacking; compared to other utilities, like Central Maine Power, Bangor Hydro had done practically nothing in terms of conservation. So the PUC fined them about a half a million dollars and told them to develop conservation programs. They said, "State law says you have to institute conservation measures before you can build expensive new generating facilities."

When I started seriously fishing for Atlantic salmon in 1984, I really fell in love with the Penobscot River, and I think the Atlantic salmon is just a magnificent creature. After getting one on the line I was hooked myself, I think forever. I think that if the Basin Mills battle is lost, the battle for restoring the

Atlantic salmon all over New England is lost. In 1967, this was named the model restoration river for all of New England. The computer-model study that was done by the US Fish & Wildlife Service is a very reliable model. It shows that without the Basin Mills Dam there's a 72-percent probability of restoring a self-sustaining run of wild Atlantic salmon to the river. With the additional dam, that probability drops down to 39 percent. The state and federal governments are together spending over $800,000 a year on the restoration of this river. Certainly, if there's only a 39-percent chance of restoring the river, I can't see the federal government continuing to put down the major share of that expense. If the model restoration river, the premier Atlantic salmon river, in the country goes down the tubes because of the Basin Mills Dam, then I can see the Fish & Wildlife Service pulling out of the other rivers like the Merrimack and the Connecticut.

On the Merrimack River the Sewall's Falls Dam was proposed, and that would have destroyed that Atlantic salmon restoration program. The US Fish & Wildlife Service strongly opposed the dam there. The agency they're a part of is the Department of Interior. The Secretary of the Interior at the time—I believe Don Hodel—got a letter signed by the Congressional delegation of each state in New England backing up the Fish & Wildlife Service and saying trapping and trucking (to carry fish around the dam, which naturally would be in the way of their migration), was no restoration. The salmon should be able to swim up the river on their own, through fishways. FERC was going to approve the license at that point, but the Secretary of the Interior stepped in. This was about 1986. The Secretary of the Interior, Donald Hodel, shot the project down.

The Environmental Protection Agency has gotten involved in this fight, and it's only the second time the EPA has been involved in a hydroelectric fight in New England. The first was the Big A Dam. The EPA has told FERC that FERC has to do an environmental impact statement on Basin Mills, and that the EPA will be monitoring and checking on the EIS as it is being done, and will comment on it when it is completed. I expect that, at that time, the EPA could step in—after the environmental impact statement shows the negative impacts and the adverse impacts on the Atlantic salmon restoration. I expect the EPA could step in with an advance veto. They have the power to do that.

Bangor Hydro spokesmen like to refer to the opposition as "a bunch of disgruntled fishermen." Most fishermen are environmentalists and conservationists, but we have many other organizations besides fishing organizations involved. On a national level, we have the Sierra Club Legal Defense Fund. Trout Unlimited, American Rivers and the Atlantic Salmon Federation (US) have filed as intervenors. Some people who never fish for salmon come down here to eat lunch during the salmon run to watch the salmon jump and roll.

Groups that are normally in favor of hydropower, but that are made up of environmentalists, are opposing Basin Mills. The Maine Audubon Society favors hydropower because they consider it a renewable resource as opposed to oil and coal, but they're opposed to Basin Mills because they feel this is an inappropriate dam. The Sportsman's Alliance of Maine and the Natural Resources Council of Maine feel it's inappropriate and are opposing it. So it's not just the fishermen.

But recreational fishing is important. The Federal Power Act and the Electrical Consumer Protection Act both say when FERC judges whether a dam should be built, the recreational fishery is just as important as the power the dam will produce.

You have to fight something like the dam very intelligently. The biggest thing originally was to organize the Coalition—get all these different groups on our side. That meant traveling around the state, giving presentations to people in Maine Audubon, the Natural Resources Council and similar groups. Of course Bangor Hydro is giving presentations too. But we won in organizing the groups and getting them behind us. Through a lot of work from a lot of people in the Coalition, we got the National Marine Fisheries Service on our side, and the EPA. I was told by the wetlands coordinator for the Army Corps of Engineers, Bill Lawless, in Waltham, Massachusetts—if the license goes through Bangor Hydro will have to get a permit from them as well—that when Bangor Hydro comes to the Corps of Engineers and asks for a permit for Basin Mills, the Corps consults with their sister agencies. Their sister agencies are the US Fish & Wildlife Service, the Environmental Protection Agency and the National Marine Fisheries Service. According to Bill Lawless, they speak with one voice as to whether Basin Mills should be built or not. He personally told me that he thought major dams in this country are very difficult to get through any more. He thought it was practically impossible that Bangor Hydro, on a national level, would be able to get this dam through.

The big thing for us right now is to stop this dam on a state level. Bangor Hydro has to get a permit from the PUC, which I think they'll get within a year or two. Then they have to go to the Department of Environmental Protection, and they've already started that process. From the DEP, they have to get two permits: one is the Maine Waterways Development and Conservation Act permit; the other is the water quality certification, showing the dam would not degrade the quality of the water in the Penobscot. We expect those hearings to begin as early as January on C-Station on the Veazie Dam, and next spring and summer on Basin Mills. If we can stop them from getting those permits—especially the water quality certification—FERC will not even look at the application. In effect, the Basin Mills project would be dead if the state DEP turns Bangor Hydro down on those permits.

At this point I'm optimistic, but it's a hard fight. Bangor Hydro has a lot of money and they've put out a lot of propaganda. They've been working with certain legislators, like Senate President Charles Pray, in state government, and they feel that legislation is on the state books that allows them to build major dams on the main stems of rivers. The state legislature in effect says they can, but that doesn't mean that they should, or that it's right, especially on the major salmon restoration river, into which the state is putting all kinds of money to restore.

You have a difference of opinion between the governor and the legislature. The governor is saying, on a restoration river that is not as significant (the Kennebec), that we should tear out a dam and make room for Atlantic salmon and other anadromous fish to spawn. Richard Silkman, the head of the state Planning Office and who speaks for the governor, said in an article I read not too long ago that the governor's taken a position that the recreational fishery on the Kennebec, and the economic spin-off, is more important to the state of Maine than the power produced by the Edwards Dam. If that's the case, you can say the same thing here. Why would you tear out a dam on a lesser restoration river and then allow a new dam to be built on your model restoration river? It just doesn't make sense.

It seems that the leadership of the state legislature, John Martin and Charles Pray, is on the side of the 1983 rivers bill that was passed saying, for instance, that on the Penobscot above the Veazie Dam dams can be built. Nineteen eighty-three was a long time ago, and I think the environmental attitude in Maine has changed since then, and I think people consider other renewable resources more important than hydroelectricity. I think maybe it's time the Rivers Act was updated to include, besides wild Downeast salmon rivers, not building hydro facilities on our main restoration rivers. The US Fish & Wildlife Service has designated in their EIS for New England on the restoration of Atlantic salmon—besides the Penobscot—the Kennebec and Androscoggin rivers. Those recreational fisheries and the economic spin-offs could bring millions of dollars into the state.

I was at a meeting between Bangor Hydro and the Department of Environmental Protection three or four months ago. Dana Mirch, the hydro coordinator for DEP, told Bangor Hydro at that meeting that the state Planning Office had done a study of energy needs for Maine for the next 30 years. They came up with a figure of a necessary increase in generating capacity of 1 percent per year. That 1 percent per year could easily be taken care of by energy-efficiency measures. In effect, it's saying energy needs are flat for the next 30 years.

In 1990, Congress excluded hydropower from what they recommended in the future as renewable energy sources to replace oil. They listed solar and wind power and organic fuels (wood-fired plants, for example). The Union of Concerned Scientists put on a presentation, I believe to Congress, that was in the paper the other day, and they said if we want to eliminate fossil fuels like oil and coal, the way to go is through renewables like solar, wind and geothermal power. With those three, they said, we could eliminate half of the fossil fuels used in the United States.

Bangor Hydro is thinking like they thought back in the 1950s. You need more power? You build another dam. They're not taking into account the fact that the worst thing you can do to a river, other than dump pollutants directly into it, is dam it. The entire ecosystem is destroyed.

Stocking Smolts at Eddington

Ed Baum

Ed Baum was born and raised in western Maine, and is a graduate of the University of Maine, with a degree in wildlife management. After two years as a fisheries biologist in Rhode Island, he joined the Atlantic Salmon Commission. He is currently senior biologist and program coordinator with the Commission.

I started with the Atlantic Salmon Commission 23 years ago. Sometimes I long for the good old days, when all we had to deal with was the fish. We were talking about that coming back from a meeting yesterday. We'd just go out, do our field work, and we didn't have to operate in a goldfish bowl the way we do now. Today it's meetings, meetings, meetings and "paper" salmon!

It was interesting the first few years—we put that trap in at the Bangor Dam in 1969. There was nobody on the river and the river was dead—not many fish at all. The year we finally reached 500 fish, sometime in the mid-'70s, we had a big party. Then the Penobscot Club got revitalized. They cleared the road, rebuilt the camp and let more members in and the Penobscot program was off and running.

In the old days we'd tend the trap, take fish to the hatchery. Now it's meetings, with Bangor Hydro or CMP [Central Maine Power] or Maine Public Service or some other state or federal agency. Our program has expanded from the Upper St. John in Aroostook County to the Saco in southern Maine. We have cooperative programs with the Canadians on border waters, and cooperative programs with people in southern New England—to compare what's going on in the Merrimack and the Connecticut to what we're doing on the Penobscot. All the dam relicensing business has come about in the past 10 years—we never had to deal with that. Requests for studies, reviewing reports, it all takes time. I feel like we get more accomplished now. I think we do 10 times more work collectively than we did 20 years ago, but we're further behind than we ever were because there is so much more to do. Everything is done last-minute, and not as well as we'd like to do it. For example, all those boxes in my office are for hydro projects on the Saco and the Aroostook and the Penobscot and the Kennebec . . . some time in the next 12 months I'm going to have to read those and comment on them and meet with developers like Central Maine Power, Miller Hydro and Bangor Hydro.

A lot of people don't know what Salmon Commission biologists do. We write a lot of individual river reports and statewide reports. We also participate in New England salmon restoration planning efforts that involve the Connecticut and Merrimack rivers, anything that ties all the salmon programs together. The regional New England Atlantic Salmon Assessment Committee meets every January and assesses what's gone on the previous year throughout New England.

There is a lot of technical assistance and coordination in our program among Inland Fisheries & Wildlife, the Department of Environmental Protection, Marine Resources, the Penobscot Indian Nation.

There is a Technical Advisory Committee made up of nine scientists who advise the Salmon Commission and the US Fish & Wildlife Service on the Maine salmon restoration program. Four members are appointed by the state. They represent the Salmon Commission, Inland Fisheries & Wildlife, Marine Resources and the University of Maine. The Fish & Wildlife Service appoints four members, and the ninth is appointed by the Penobscot Indian Nation.

Commission biologists do a lot of writing. Anything from the newsletter we put out occasionally up to very technical scientific publications in fishery journals. We're asked to comment on anything that would affect the habitat or environment of Atlantic salmon: environmental impact assessments, DEP wastewater discharge permits, FERC applications for dams . . . .

Salmon Commission biologists are primarily responsible for taking the adult salmon broodstock that we need for the statewide program to federal hatcheries in Maine. There are about 600 adults we have to collect each year. Plus we're also involved in fry, parr and smolt stocking, and evaluating how well those fish survive in the river, and when they come back in too. We've collected samples of some of our wild salmon all over the state to try to evaluate the genetic diversity in our wild stock, in order to see how different each river is. We'll be comparing those to aquaculture stocks and to our hatchery stocks. Eventually we hope to develop local salmon stocks for local rivers in order to speed up the salmon restoration. Finally, on the international level, we participate in biological sampling and tagging recovery programs. We've sent people to Greenland and Newfoundland/Labrador, where we collect biological samples and retrieve tags from tagged salmon from Europe and North America. We provide scientific advice to NASCO [the North Atlantic Salmon Conservation Organization] through our participation in the International Council for the Exploration of the Sea, both working groups that meet annually in Europe and North America.

With the recent budget cuts we've had to revise our statewide program. It's unusual, because we've now got six people working on one small river [the Narraguagus study, a $200,000 project], and two people and about $80,000 to spend on all the other rivers in the state, including the Penobscot. The state didn't appropriate enough money to fund the three General Fund positions we had left. We're going to have to spend our money and get reimbursed for some of our work with the "Senator Mitchell money" to pay our salaries later in the year. Senator Mitchell had $100,000 appropriated for the Salmon Commission to use through the Anadromous Fish Conservation Act administered by the US Fish & Wildlife Service. We apply for the funds—stating what we will do for work—spend our own funds and apply for 50-percent reimbursement. Otherwise we're going to run out of money at the end of the fiscal year. One hundred thousand dollars is just going to keep us barely going for another two years. The other part, too, is that all state employees are going to have to take furlough days without pay. I had 27 days off without pay this year [1991—because of shortfalls in the Maine state budget]. When employees take their vacations, holidays, and numerous furlough days without pay, it doesn't leave much time to get the job done.

On the Penobscot River, we need to get a minimum of 3,000 spawning females upriver to make full use of the habitat. We've only reached an average of 21 percent of that goal each year since 1977. In 1991 it was only around 15 percent of the goal, and that's why this year we're trying to reduce the sport harvest—to allow more females upriver to spawn. In 1986 we released the largest number of [returning] females from the trap at the Veazie Dam, 1,400, and five years later, in 1991, we had the largest return of wild fish [470]. We are concerned about the [downward] trend in the survival of hatchery smolts released in the Penobscot in recent years. We've had poor years here and there, but we've never had six or seven poor years in a row before. Basically since the early '80s, it's been going down, not only on the Penobscot, but on many rivers all over North America. Everyone's looking for one single reason, but it's probably not just one thing—it's a combination of many things at home and at sea. Of course at NASCO meetings, the harvesting countries say the problem is in home waters.

Global smolt survival in recent years has declined. Many Canadian rivers have exhibited the same type of low survival seen on the Penobscot and other Maine rivers. Something's going on in the ocean we don't know about that's affecting all salmon stocks, of both wild and hatchery origin.

Despite all that we know about Maine salmon, we have some questions about migration problems. For example, some tagged Penobscot salmon go to Labrador and Newfoundland, and some go to Greenland in the same year. They don't have time to go to both areas in the same year—we don't know why some go one place and some another and in what proportion. Also, there are questions about separating fish from individual European countries and those from North America that are caught in Greenland. This resulted in the formation of NASCO in February of 1984. It's an international organization that the US belongs to and it has jurisdiction over all salmon stocks north of 36 degrees latitude in the Atlantic Ocean. They have the same objectives we do: conservation, restoration, enhancement and management. In NASCO there are three different commissions. The North American one involves the US and Canada; the Northeast Atlantic is basically European countries; the third is the West Greenland commission. Each one of these commissions regulates the fisheries in its area. The three commissioners who represent the US are Alan Peterson from the National Marine Fisheries

Service in Woods Hole, Massachusetts—he's also the NASCO president now—David Egan from Connecticut, and Bill Townsend from Skowhegan, Maine. These three people represent the United States in all the international negotiations that affect our fish. Very influential people in very responsible positions.

A lot of European stocks go to Greenland too, like ours, to feed, but the ocean temperatures are nowhere near as harsh for the European stocks —they're much milder. There's a tremendous difference in temperature from Maine to Greenland, and from Norway to Greenland the change is nowhere near as great. It's speculated that that's why European salmon smolt survival is so much higher than ours. It's not uncommon in some European rivers to have a 10- or 20-percent return of smolts.

In Canada, the commercial catch has gone down from 2,500 tons in 1982 to less than 1,000. Greenland was way up around 2,600 metric tons in the early 1970s, and is down to less than 500 in recent years. So the Canadians and Greenlanders say, "We've reduced our catches—we're not the cause of your poor returns."

We've been studying the percentage of wild fish in the run as compared to the number of hatchery fish. The cycles are five years long, and we started with virtually no fish at all in the late 1960s—they were completely destroyed by dams and pollution. It takes a lot of five-year life cycles to build up a salmon run. Now we're beginning to see a lot of progress. The numbers of hatchery fish over the last five years are now up to an average of about 3,000 fish annually. The wild fish are what we're really interested in. The numbers have roughly doubled in each of the last three generations to an average run of about 300 wild fish during the last five years. We're really encouraged by this and anxious to see if this trend continues for another three or four generations.

Regulation of the sport harvest is another significant problem. We know we're not meeting spawning escapement on the Penobscot. Even with a one-fish limit, we're finding exploitation rates of around 9 to 10 percent on hatchery fish and, interestingly enough, with the wild fish coming in earlier in Maine, we're finding nearly double that—18 to 20 percent—on the wild fish. And those are the ones, of course, that we want to get upriver and to have spawn. How to regulate the harvest and still maintain angling opportunity and allow anglers to catch fish and not have to release them all is something we're wrestling with. We have three regulation options we're considering for 1992 to reduce sport harvest statewide and the Commission will be seeking the opinions of anglers at public hearings. I'm sure there will be some changes made, but it depends on what comes out at the public hearings, as well as the written comments we receive. Reducing the harvest to allow more spawning escapement on other Maine rivers is also very important. For example, right now you can catch five large fish Downeast on any one of those rivers.

Many of the anglers have said, "It's not worth fishing if I can't keep anything but grilse," and a lot have said they won't come back and fish if Maine adopts a grilse-only law. But with the little bit of testimony we've received so far they seem to say that as long as we give them something to fish for—if, for example, they catch a big, trophy fish they can keep it—they have a reason to come back and buy a license.

The Downeast comments were, "Well, we went from 10 fish to five, we went from two a day to one, and the runs are worse than they ever were, so why should we be restricted more? It's not the anglers' fault." Everyone blames everyone or something else—seals, cormorants, commercial fishing, biologists, dams on the rivers, we don't stock enough. Everyone's got an answer, but of course it's never the anglers. My opinion is that they can either become part of the solution or continue to be part of the problem.

You can't have it both ways. Anglers can't say the runs are poor, the fishing's terrible, and then continue to ask to kill five fish a year. Finally, I think a lot of people are beginning to realize now that cutting back is the best thing we can do. [For 1992 at least, the regulation was changed to a one-fish limit—for the season—statewide.]

Building C-Station at the Veazie Dam could affect fish passage and angling; overall it could be better for fish passage in the long run—certainly better than the conditions we have right now. Bangor

Hydro is currently spilling a lot of water over the dam in May and June. If they put C-Station in there will be a lot less spill and the flow will be concentrated. The fish will find the new fishway and get upriver much sooner than they do now. The new plant will change the flows along the Eddington side—it may be worse, or it might be better for fishing. It could create angling up at the Basin Mills site or some other upriver area that we don't have now, because fish will move up earlier and faster. People could be fishing in Orono in May, as opposed to Veazie and Eddington under current conditions. However, I wouldn't advocate the construction of a new powerhouse for those reasons. What we gain in upstream passage may be lost due to smolt mortality when juvenile salmon move downstream past C-Station.

At Milford, the DEP expects to be issuing a permit soon for the relicensing of that dam, hopefully specifying "state-of-the-art" upstream and downstream passage—that's what we've asked for. Additionally, James River is proposing to rebuild the Great Works Dam, and they're offering new upstream and downstream passage there—10 years earlier than they normally would have to. Within five to 10 years we should have really good upstream and downstream fish passage facilities at most dams on the Penobscot. This is one of the reasons that I am so optimistic about the future of the Penobscot River restoration program.

Schoolhouse Pool on the Mattawamkeag

way you're going to vote." So in the case of politics, the three-man commission that was supposed to be an independent commission really wasn't an independent commission. Only one of the three was.

The other instance was in the rebuilding of the Bangor Dam. I was against the rebuilding, and the other two members didn't take a stand because they were told by the governor not to. I don't know if the same politics prevail today, with three public members and two administration members.

I think some of the fishermen wanted three public members and two administration members so there could be a 3-2 vote. But that can also come back to haunt you when the public members are geographically oriented or have other vested interests. I was a political appointee to a certain extent myself, because I was a good friend of Brennan and was instrumental in getting some people elected in the Democratic party locally and on a state level. Nevertheless, I was a salmon fisherman and a person concerned with salmon. The two public members that would have replaced me now that I'm out are salmon fishermen, but only one of them has fished on this river and he only fishes once or twice a year. Joe Sewall, and Rick Warren from the *Bangor Daily News*, have a very nice lodge and own an awful lot of water on the Upsalquitch River in New Brunswick, so they aren't about to fish here anyway.

When there was a stand to be taken against the Bangor Dam, Joe Sewall definitely put his power and prestige behind the killing of the dam. Make no mistake about that. But Joe is just too busy, and he'll tell you himself he's too busy, to commit himself to the Atlantic Salmon Commission.

One of the problems is that the money Bangor Hydro is spending on trying to get Basin Mills built is from rate hikes, so it's not as if they're throwing away their own money. The only thing that might happen is public and financial pressure. The stockholders might say, "That's it. Get the hell out of Basin Mills, because we're not going to have the cost of that taken out of our dividends."

You have to remember how Bangor Hydro has been run for years. Bob Haskell ran his company just like he wanted. Haskell was governor for five days, and so we call him Governor Haskell. Thirty years ago, when Bob Haskell and George Everett, who used to have controlling interest in what's now

Fleet Bank and Merrill Bank, got together and said, "This is what's going to happen in eastern Maine and Bangor," that's what happened. There was no questioning that. You didn't get loans, you didn't get anything unless those gentleman knew about it. The times they are a-changin', and they have changed.

But the PUC fining [of Bangor Hydro for mismanagement] the other day may indeed force some stockholders to go ahead and say something. It's going to do something to the board, because the people on the board—and most are major stockholders in the company—are going to say, "Hey, you aren't going to fine me one half of one percent of my profits." And that's all it was, incidentally. "If it's going to cost me this much to pursue a Basin Mills that won't take place, then drop the goddamn thing right now and stop the cost right now."

The Public Utility Regulatory Policies Act that was passed [in 1978] after the artificial price hikes by OPEC was meant to encourage energy production here so we would be energy-independent. It wasn't meant to be hydropower alone, but the guys with Swift River, who wanted to rebuild Bangor Dam, took advantage of it by selling limited partnerships to people who wanted the tax break that went along with the same law. There was no benevolence on their part to produce power so we would be energy-independent. That had nothing to do with it. They were getting a 21-percent break on their taxes. Then they go ahead with rapid acceleration depreciation and depreciate the whole thing in three years, then sell it and not lose any money. It was a hell of a tax deal.

It was also a hell of a battle. It was a marshalling of not just the entire state of Maine but people all over the United States. They put pressure on their legislatures and said building it couldn't be justified.

We—the ASC, the fishermen, the salmon clubs—were on top of it from the beginning. LD 1800, which was passed in the mid-'70s, made sure that Bangor Hydro could not end up transferring the rights to rebuild to anybody else. There was no way Swift River was going to have that thing all by itself.

Credit was never given, and it should have been to a certain degree, to Howard Trotsky for 2048. It said no new dams could be built below the Veazie Dam. He wrote the bill, and then John Martin,

Ken MacLeod, Joe Sewall and Dick Ruhlin pushed the legislative council to make it an emergency measure and get it through in the last second. We still can't do anything about Bangor Hydro rebuilding Bangor Dam, but they could never get the bonds to rebuild it, so they were willing to let Swift River do it for them. LD 1800 kept them from transferring the rights to do that.

Trotsky was the one who had the brains to say, "Hey, let's put a bill in that says they can't put a dam in tidewater." But he was a little too wild to let into the legislative council. Ken MacLeod and Dick Ruhlin and those guys said, "For God's sake, don't let him in. We'll take care of it."

The nature of the threats really hasn't changed much. We've always had our fishway problems, and they're still with us. Downstream passage is still a problem. No one really knows what works. Al Meister used to say that one out of every 10 springs is a dry spring, and that kills your downstream passage. But most of the time, if you get enough water, the fish can go down over [the top of the dam] anyway. You're always faced with the money problem. The Salmon Commission's always been understaffed, never had the time to put in the scientific research that those people were trained to do. There are too many menial tasks that those guys have to do. When you have a biologist tending a trap, that's a poor allocation of the resource. He should be doing research, but we've never been able to do those things.

Even the management programs never said we intended to build a self-sustaining run of salmon to the Penobscot. The evaluation was that the best we would have would be 50/50. It's because of the dams that are already there—access and egress. Not enough of the fish are going to get to the spawning grounds. The figure of 10-percent loss at each dam is tossed around. That's why we had to truck the fish initially. You had to have the maximum number of fish in the spawning areas.

I'm not even optimistic about reaching that 50/50 level. But I still don't know the wild fish figures—whether or not they're multiplying at an arithmetic rate or a geometric rate. They're certainly not where we expected them to be by 1991. They still don't know whether the smolt program is indeed the right thing to do, they still haven't done the studies necessary to decide whether parr stockings are the best way to do it. They still haven't done enough scientific research in the entire world to know whether fry stocking is the best way to go. Or maybe streamside incubation. We've got the expertise in Maine to do it, but we don't have the time or the money. You just can't have people the caliber of Ed Baum and Ken Belan and Norm Dube, Al Meister when he was there, doing meaningless things, shuffling papers.

The problem with money is that the Atlantic Sea-Run Salmon Commission is an independent body, but all the regulations come under Title XII, which is Marine Resources. The money comes out of the Marine Resources budget, but it's given over to Inland Fisheries & Wildlife, which disperses the funds. It's a mess. The federal contribution is simpler. They sit there and raise 750,000 smolts a year in federal hatcheries, give them to us and say, "Do what you want with them." They don't tell us where to put them; they even help us put them where we want to put them. It's amazing.

We may not have done the Downeast rivers any favors by giving them Penobscot River smolts. We could have killed their runs. That's why we always at least held back on the Machias River and never put any smolts there. But local people complained that everyone else was getting smolts and they weren't, so they're getting them now.

The Ducktrap gets genetic slop from fish that are coming back to the Penobscot but end up there instead. You actually end up with more fish going into the Ducktrap River in the fall than the Ducktrap can hold. That's why even the wardens never got too excited if someone took an extra fish down there, or snagged one. But that's good genetic slop, a natural stirring of the gene pool.

I think it's absolutely crazy for the Downeast rivers to keep a five-salmon limit, instead of going to one [there is now a one-fish limit, statewide]. Stop and think. The Miramichi River, with its fantastic runs, had a problem, and it was decided that no adult fish would be killed that year. And yet the fragile Dennys, East Machias, Machias, Narraguagus, Sheepscot and Ducktrap rivers have limits of five. It's crazy. Even on the great rivers like the Restigouche—one salmon and that's it.

We always thought that no obstacles is what would save the Machias River, when we removed that last dam at Whitneyville. That was the biggest disappointment we ever had.

One very, very positive thing that I've seen is a great concern on the part of the salmon fishermen themselves for conservation. When we went to a 10-fish limit, the majority approved of it; when we went to five fish, they approved it; and they almost demanded the one fish. The salmon fishermen on this river, and all the rivers, deserve a lot of credit for releasing their fish, and being genuine about it, and almost being overly concerned that somebody does it and does it right. Just like when we just watched that guy bring that fish in, you said, "Jeez, that's great, he got that fish in fast, unhooked it fast and got it back in the water." If someone's taking too long with a fish, someone else on the river will say, "Why doesn't he get that fish in? What the hell's he doing? That fish is gonna die." They're starting to take each fish to heart. When the first salmon came back to the Penobscot in 1965, everyone just wanted to kill all they could kill, hold them up and say, "Look what I got." Not any more. I think it will get into the Downeast rivers the more the fishermen there see that the state and the Atlantic Salmon Commission are genuinely concerned about them.

We had to do it because it's a model river, but we did put too much attention on the Penobscot River from '72 to '80, and the Downeast fishermen resented that, even though 99 percent of the salmon fishing was done on the Penobscot. The ASC is going to put more focus on those other rivers in response to the river management programs. The fishermen there may start picking up the attitude of letting those salmon go.

The response of these guys when something goes wrong is something. When they started to think of rebuilding the Bangor Dam, Woody Higgins, then the president of the Veazie Club, stood up at the first meeting there and said, "We've got to do something about this." Right off the bat, it started. The same has happened with Basin Mills.

It's all so archaic. We're still producing electricity the way we did 100 years ago. It's crazy. There would already be better ways if someone said, "That's it, boys. When your 40 years are up, you're going to take the dam out."

Penobscot Story

Jerry Marancik

Jerry is Maine Fishery Program coordinator for the US Fish & Wildlife Service, and acts as the liaison between the Fish & Wildlife Service and other agencies involved in the anadromous fish restoration program for species such as Atlantic salmon, American shad, striped bass and herring.

The National Marine Fisheries Service is contributing quite a bit toward funding studies on the Downeast rivers and the Penobscot, and the US Fish & Wildlife Service has put in a $100,000 matching grant this year [1992]. I think the state can match that over two years, at $50,000 a year.

We own the hatcheries and the land, so even when budgets are cut we still have the same capacity to raise fish. It's a matter of utilizing the resources. We talked about not stocking the Saco or the St. Croix, and people said, "How can you do that? You have the same number of fish." That's true. But [with reduced funds] we don't have the same number of people to stock the fish, do follow-up studies, collect returning adults, etc.

I guess I'm opposed—and I think most people are—to just putting fish in a river and walking away. There should be some monitoring of return when we stock fry, there should be some monitoring of survival. We try to learn from everything we do. If we put fish in any river, there's a commitment of time, money and people to follow up on it.

Stocking fry is very labor-intensive, and we can use volunteer help. We've never really been geared up to get large numbers of fry out efficiently. We've always had maybe 200,000 or 250,000 fry;

we're gearing up with equipment for stocking one million fry in 1992 now, and we're going to be lining up volunteer help.

We often feel like spiders, getting pulled in every direction, because we are involved in 15 rivers from the Aroostook south to the Saco and everywhere in between. We're not stocking enough fish in any one river, I can guarantee that. We're not stocking enough to satisfy any of our own requirements, or any of the anglers'. There are really only two hatcheries in Maine, and they're both federal hatcheries: The Craig Brook hatchery in East Orland, and the Green Lake hatchery in Ellsworth. There just aren't enough fish to go around.

We raise approximately 100,000 fry a year at the Craig Brook hatchery, about 50,000 to 75,000 parr a year and 150,000 to 175,000 smolts. Green Lake is a more modern facility, started in 1971. It's a larger facility, raising around 600,000 smolts, 200,000 fry and 200,000 parr a year.

Each year we release about 600,000 to 800,000 smolts [500,000 to 600,000 in the Penobscot]. In 1983 or '84 we went from a two-year smolt program to a one-year smolt program where, instead of holding the fish for two years in the hatchery, we incubate the eggs and hold the sac fry in warmer spring water. We get them accelerated so they only have to spend one year in the hatchery. By doing that we're able to increase the number of smolts we produce. There's also been an increase in fry production, and that's an attempt to improve the genetic composition of the fish. I think we'll see fry stocking increase as time goes on. There will be more emphasis on fry production because smolts resulting from fry stocking or natural reproduction have undergone one process of natural selection, are more genetically adapted to the river environment and survive better than hatchery smolts.

We started from virtually no stock on the Penobscot, so we used smolts to develop a run and broodstock. That worked—we're getting 2,000 to 4,000 fish back now. Now those hatchery smolts have kind of just leveled off. You put in so many, you get so many back. What we're trying to do now is what I refer to as Phase II, and that's go to more fry stocking. We'll probably take more broodstock and eventually release maybe 5,000,000 fry. Again, you go through the genetic selection, and theoretically these fish will be more suited to the Penobscot. Instead of getting fish back now we're going toward restoration [of a wild run], and that's exciting.

Records show that Atlantic salmon were abundant in colonial New England, with an estimated 300,000-plus adults returning to 28 rivers. I recently read one estimate that put the number as high as 1.1 million. However, dams, overfishing and pollution eliminated most of the runs by the mid-1800s, and even as early as the late 1700s on the Connecticut River. In the latter part of the 1800s, a group of farsighted men saw the problem with Atlantic salmon and had the foresight to attempt a restoration program. This is significant to the Fish & Wildlife Service because it was the beginning of the Craig Brook federal hatchery, which started in 1871 and is the oldest salmon hatchery in the country. It was started by a fellow by the name of Charles Atkins, who really was the father of Atlantic salmon culture. At the time he was a commissioner of the state of Maine, and then he jumped ship and became the commissioner of what I believe was the Commission of Fish & Fisheries [a federal agency]. They actually built wooden fishways around dams, stocked fry and parr in the Penobscot, and in the Hudson and in rivers all around the East Coast. They used wagons and trains, with milk cans, to transport the fish. But things like floods wiped out the fishways, there was still pollution, still overfishing. There were no laws to require fishways, so when they washed away they weren't rebuilt. But they *were* quite successful in bringing a lot of these fish back.

Atkins kept detailed journals, and it's interesting because we have many of the same problems he did 100 years ago. I'd like somebody from the University [of Maine], somebody's who's interested in biology and history, to summarize what's in there. Atkins was really progressive. He was really a smart man. He was the first one to tag salmon. He tagged kelts, with little aluminum bands around their tails, and they came back with them on. Crude, but it worked. I'm sure we're doing experiments now that he probably did 100 years ago. No one has the time to find the answers in his journals.

I worked at the Berlin, New Hampshire, hatchery for two years, and it was the same type of thing. From day one they kept a daily log. It had the details of what everybody did, they went through the

CCC and WPA days: "Seventy-five workmen showed up today to dig a big pond. We sent them home because it's 75 below zero . . . ." We're not going to have records like that for future generations.

Craig Brook started as a salmon hatchery, which shows the foresight of those guys, because they realized salmon stocks were declining rapidly. They stocked mostly fry; they didn't have the technology to raise them to smolts. They fed them maggots—they'd hang rotten meat in the fly house, the flies would lay eggs, somebody would collect the maggots and they'd feed them to the fish.

They raised a good number of trout here as well, and did research. There's circumstantial evidence that they did cancer research here 50 or 60 years ago. But it's always been a salmon hatchery—just the major mission changed. They started raising salmon again when the Atlantic Salmon Commission started in 1948. But we still have the same problems: overfishing, pollution, dams.

We got excited during the Craig Brook rededication in 1989 [marking the 100th anniversary of the hatchery's inclusion in the federal system]. Atkins' records were up in the hatchery attic. We got them out of their boxes and brought them down here. Some of the better records are in the National Archives. They're preserved, they're cataloged, but they're totally inaccessible. If you really want to do a big research project, you can go down there and spend a week or so doing research.

What we're trying to do here is rebuild what nature took 10,000 years to build, starting from scratch, and we get upset if it doesn't happen in 10, 20, even 40 years. I think we've done really well with the mediocre fish passage we have, the lack of downstream fish passage. We don't stock anywhere near the number of fish we need each year. In 1991, for instance, we should have had about 1.2 million smolts in the Penobscot, and we only had about 500,000. I think we've done really well with one hand tied behind our backs. If we could get real state-of-the-art upstream and downstream fish passage and five years with six, seven, eight thousand fish upstream to spawn, it would really take off. The five-year life cycle hurts—the eggs we take this fall [1992] won't come back as adults until 1997. That's a long time to wait, and a lot of things change.

When you look at the Penobscot and what it was, with pollution and other problems . . . the value of it now that it is being cleaned up, and what it could be to Bangor, Brewer, Orono, Old Town. And the Atlantic salmon—which is in itself a miracle when you think of its life cycle—to think that it can survive again in the Penobscot River. Just knowing that it's there is worth something to me. I hope other people feel that way. I'll never see an African elephant, but I'd sure feel bad if they became extinct. So I would like to think that people would support the program even if they can't catch and kill a fish, for instance.

One of the other problems, and it's our own fault, is that the salmon program is looked at as an elitist program where these high-powered yuppies from Connecticut come up and fish, and that's why we're putting fish in the river. It's not true. We're restoring the salmon for the sake of the fact that they should be allowed to survive. The salmon should exist here, the same as the peregrine falcon or the bald eagle or the whooping crane. We have the technology now, we have the incentive to restore the salmon. That's why we're restoring them. If there is some angling effort or enjoyment that some can get from that, that's fine, but that's not why we're doing it.

We should be talking to the people in Rumford [Maine], for instance, and saying, "I know you don't live on the Penobscot, but these fish are for you too. You're helping pay for them." I think the program is looked at as elitist around the state, and that's our own fault, because we haven't presented it to the general public as much as we should have. I think that's why people say, "Well, if you don't allow any harvest," (which is not true because people could still take grilse) "we won't fish anymore." If that's our only support, then we're in trouble.

I'd be leery about saying this in public because it would be misconstrued, but as far as money goes, anglers contribute only about $25,000 to the program a year, when the Fish & Wildlife Service is probably putting in $1,000,000 or $1.1 million a year. And then the state is putting in some. That would be misinterpreted. But the fishermen provide moral support, and that $25,000 pays someone's salary in the Salmon Commission. We need the support of the anglers, but we also need, and deserve, the support of the general public.

On the Merrimack and the Connecticut rivers, there is no harvest of salmon allowed. There probably should never have been any harvest allowed in Maine from 1948 when the Atlantic Salmon Commission started. Yet down there in southern New England they get so much publicity. This year [1991] they got maybe 200 fish back, and it made front page in the news. Here, if we get 2,000 fish back in the Penobscot, everyone wonders what went wrong.

I'm beginning to look at the salmon program like a big tree. You have the salmon restoration program at the base, and that seems basic enough. But then as you get up into the branches, it starts getting a little more complicated. Then you branch out even more and even more, because there are so many agencies, so many groups involved, and the nature of the animal itself is so complicated. Right now the runs are down everywhere. Not just the Penobscot, but also Downeast, and in Canada . . . . We try to look for easy solutions. People say, "Why are the runs down? We have to stock more fish because the runs are down." But that may not be the answer.

I was at a meeting a few weeks ago, and a guy from the National Marine Fisheries Service was giving a presentation, looking at ocean temperatures between the Gulf of Maine up to Greenland. The ideal temperature for the salmon is shrinking [in geographical distribution], so the habitat in the ocean is smaller than it was, say, 10 years ago, which certainly may affect the food. It may affect a lot of things. There's nothing we can do about that. It's probably something that just happens every now and then, it's probably happened for thousands and thousands of years.

Virtually every population has a cycle—acorns, bunny rabbits. But we can't tolerate it if it's the salmon. We've got to fix it or find out what's wrong and what we're doing wrong. Sometimes we may be doing more harm than good. Say Downeast. The runs are down, so we start throwing a bunch of Penobscot stocks in there to fix the problem, but in the long run we may be doing more harm than good. If we just protected them and let them alone, they might come back and do fine.

I think the days of just being a sportsman, or just being a private citizen, are over. You're going to have to get involved in the environment somewhere, even if it's just recycling. People are going to have to understand the problems. And yeah, if you're a fly fisherman or a saltwater fisherman, you should understand that there are problems out there, and if you want to enjoy it just maybe you can do something to help.

The bottom line is that we all want to restore salmon—it's just a matter of why and how, and that is sometimes where we lock horns. The program is cooperative, so we do work together. We [the federal government] fund and run the hatcheries, the state has management authority. Some people don't realize that. Like angling regulations—that's a state function, we have nothing to say about that. What I'm proudest of is the way it's working. We do work together, and it does mesh. It takes some discussion, and there is some disagreement, but we never let it really get in the way of the program.

The Salmon Commission staff makes recommendations on where the fish go—smolt, fry and parr—and then it's presented to the Technical Advisory Committee, and we hash it out. The TAC is made up of four members from the federal government, four from the state government and one from the Penobscot Indian Nation. The final recommendations go from the Technical Advisory Committee to the Commission. The final allocations are agreed upon by Bill Vail—the Commissioner [of the Maine Department of Inland Fisheries & Wildlife]—and our [US Fish & Wildlife] Regional Director.

We're involved in the FERC process. Gordon Russell [of the US Fish & Wildlife Service] in Orono is the lead on that. Again, it's a cooperative thing. Quite often, before we comment on anything or take a position, we make sure that the Salmon Commission, IF&W, Department of Marine Resources, Fish & Wildlife Service, National Marine Fisheries Service discuss the issues. Again, sometimes there are differences, but generally they're very small. We're very rarely at odds over anything because the welfare of the resource comes first.

The farther you transport fish away from their "home" river, the worse they do. We brought fish in the 1960s from the Miramichi River in Canada and put them in the Penobscot, and they did terribly. But we brought fish from Downeast and put them in the Penobscot and they did better.

Out West, particularly in the Columbia River, they had good runs of Pacific salmon, and they put some hatchery fish in and that was fine. Then they put some more in, and that was all right. Then they put 12,000,000 in, and shortly after that all the runs went downhill. We suspect that the hatchery stocks spawned with the wild stocks and diluted the suitability of the fish. Hatchery stocks are not as suitable. It's a numbers game, where you can put some fish in and the wild stocks will do fine, but if you overwhelm them with hatchery fish then you may be doing serious harm to the wild stock. They have looked back and shown time and time again that hatchery fish are less suitable than wild fish, and if they hybridize it results in a genetically less suitable fish.

Then you can say there are no stocks left Downeast. We hope to find that out shortly, because we're looking at the DNA composition of those fish. I don't know if that will give us the answer we need, but my feeling is that we should be on the conservative side. Those fish may be part Downeast, they may be pure Penobscot, or they may be pure Downeast, pure Penobscot and some hybridized, which is what I suspect has happened. But even those hybrids are better than bagging the whole thing. We're still better off perpetuating those fish and letting them weed themselves out genetically. It's going to be interesting to see what those DNA analyses show.

We're in the process now of converting the Craig Brook hatchery from Penobscot smolt production into a multiple broodstock facility, where we'll take stocks from individual rivers and raise them to maturity, then put the progeny from those fish back into the rivers rather than having one stock for all of Maine. The purpose of that is to keep the gene pools [of different rivers] intact. Unless the DNA analyses show that those fish Downeast are 100-percent pure Penobscot, exactly like we have in the hatchery, I think we should just proceed with what we're doing and try to preserve them.

They raised fish here at the hatchery from the Machias, the Narraguagus, the Orland River, the Miramichi, several other places. That was kind of the nucleus for the Penobscot stock. The runs never totally disappeared, but they were so low you might as well consider them to have disappeared.

Malcolm Coulter

Mal Coulter received his master's from the University of Maine and his PhD, in forestry and environmental science, from Syracuse. He spent his career in the fish & wildlife departments of Connecticut and Vermont, and at the University of Maine, in the Maine Cooperative Wildlife Research Unit and the College of Forest Resources.

I came back to Maine in '48 as assistant leader in the cooperative research unit at the University of Maine. It was a cooperative thing between the Fish & Wildlife Service, the state and the university. For 10 years I wasn't involved in teaching, I was in the field all over the state doing field research.

At that time we had four graduate assistants—people working toward advanced degrees. I spent a lot of time with each one of them. Those early years were great years because you were in the field. Later I began to get into teaching, and I finally got to the point where I had to make a decision whether I was going to stay here or go with the federal government. I elected to stay in Maine, serving as associate director of the school of Forest Resources for a while, and when it became a college I became chairman of the department of wildlife. I retired as chairman of the department of wildlife, which had nothing to do with fish. Fish were split, over in zoology.

Bucky Owen replaced me. We hired him in 1968. I took him on his first trip, when he got his first salmon. I was very happy to see him take over the department—he's doing a great job. This past year the wildlife department of the University of Maine took in more money from outside research grants and so forth than any other department except for chemical engineering, which gets a lot of money from the pulp paper industry.

I got interested in salmon because, I think in 1950, Bill Peppard and I went to the Bartibog River in eastern New Brunswick fishing for spring salmon. It was a situation similar to the Penobscot in the spring, with the fish in an estuary where they drifted back and forth with the tide. By the time we got there the fish were feeding on smelts; they'd turned around after the spawn and were very, very active. I went there for two or three years, and that's where I got introduced to salmon. Then we began fishing the Narraguagus and other places.

I started going to some of the early meetings of the Salmon Commission, or occasional ones. I knew Horace Bond quite well. Part of his career he sold grain for the Red Rose Feed Company, and we bought a lot of grain for trapping wild ducks for banding and so forth. He had a camp down to Beech Hill Pond, and for some reason—I was just a young fellow—I'd get invited down there with him and some of his cronies, to little stag parties where we'd play cards after dinner and after Horace had told his stories. And he had many stories, and he could recite poetry from now until Hell froze over. He was a great guy. He was really dedicated to salmon.

I took my first fish from the Penobscot with Bucky Owen from his peapod, although I've never fished here much. Leroy Hutchings netted it. And we used to fish the end of the Bangor Salmon Pool from my double-ended peapod.

The restoration has gone slowly, and I think it's gone slower than many people anticipated, including some of the biologists and others. I think we're naturally impatient. If you look at some of the old reports, they are very optimistic in the '40s and early '50s about what was going to happen. But if we look at it over a period of time, we can compare today with the '50s or the '60s, and certainly the '70s, and we've made great progress in the face of a hell of a lot of obstacles. The obstacles include not only the obstacles in the river, but obstacles that are away from the river, beyond our immediate control. I was very interested in what Alan Peterson said the other day [at a meeting of the Northern Penobscot Salmon Club], when he made the point about the program and the conditions in the local river. We all want to blame the high seas fishery, and it's clear it has an impact, but he was highlighting the importance of what goes on in the local river. Deterioration of the habitat, or how the resource is used or abused or what have you. What goes on in the local river is tremendously important.

Of course here we're up against some obstacles we'll never overcome. That doesn't mean that we won't have a successful restoration, it means that the numbers of fish that ultimately come to the river or spawn in the river won't be as great as they might have been. We have deposits of sawdust in the estuary that are 15 feet deep. They've been there for 100 years. Farther upriver there are lots of places where there are two, three, five and six feet of sawdust and bark from the old sawmills and the old log drives. When I first went upriver in the late '40s and early '50s, there were sawmills that blew their sawdust in the river. Or they blew their sawdust into piles in the floodplain when the state prohibited them from putting it in the river, and every second or third year the sawdust pile was swept away in high water. So spawning areas get covered. Siltation takes place. Tributary streams are plugged one way or the other, silted in. I guess what I'm saying is that one of the things we're facing is a loss of habitat we probably can't restore. But there is a great deal of suitable habitat. The problem with the suitable habitat seems to be that we have difficulty getting the fish to the suitable habitat, and from the suitable habitat back to sea.

Look at the 1990 figures. We had 3,700 fish in the run. If you look at the river, and this is true of most rivers, the good spawning habitat is in the headwaters. We have the East Branch and Passadumkeag Stream. The Pleasant River up through Brownville is really nice. The Mattawamkeag doesn't have as much by way of really cold water, but it has some. But these are places where you'd like to get the fish. The East Branch is probably the premier spawning and rearing part of the Penobscot River. We get less than 10 percent of the run up there.

Another aspect of this whole thing that always made me wonder, although I've never seen any information or background on it . . . this fish comes into this river with all the calories it's going to have until it goes back to sea. That's got to take it over the dams, it's got to take it up the river, it's got to get it to wherever it's going to spawn, it's got to get it through the spawning act, it's got to get it back to the sea. I can't help but believe that every time a fish has to fight through a fishway, for example . . . you just wonder what this does to the condition of the fish. What does this do to the spawning efficiency of the fish? We really don't know. We know some of them spawn, but do they not construct as many redds, or do some of them not spawn at all? These are obstacles we've placed in the way of the fish, and then we have the reverse coming down.

Go back and look at some of the old reports of the early stocking in Maine, on the fingerling stockings. When you stock fingerlings, you put these little fish that have just started feeding in these little tributaries. You put many more of them [than smolts] out, because of course there's very high mortality. They become acclimated to that, and when they go out they're predator-wise. The stocked smolt doesn't react to a cormorant, for example. If the fish has lived two years in the river, by the time he gets down here as a smolt he's avoided a lot of predators, and I really think it makes a lot of sense. And it makes sense to concentrate more effort than we're able to do now with our limited resources on habitat work in those tributaries, and protecting those tributaries.

It's really frustrating to see how much money has been raised, basically by fishermen and from fishermen—although not all of it—to fight dams and this sort of thing, when that same amount of money could have been applied to other things on the river.

I'm concerned with the article in the paper the other day. Kevin Boyle, who has been doing good work in Maine on the economic aspects of hunting and fishing, has said it might be better to keep the Edwards Dam [on the Kennebec River], if they put a fishway in. What the hell does Kevin Boyle know about fishways and their efficiency? Unfortunately lots of people think that if you have a fishway you solve the problem of fish passage. Their efficiency is 40 to 90 percent, and 90 percent is only some of the time; it isn't in high water, low water, at all seasons of the year at all times. You may get 90 percent at a good fishway under ideal conditions. But ideal conditions don't prevail often.

We've done something to the early genetic stock on the Penobscot. The problem we have here is apparent right now [late in the season, during low water and high water temperatures]. We're getting a hell of a lot of fish coming through under conditions when they provide very little sport. The early run of fish that used to show up here, the early genetic strain, they'd start catching them in April, they had good fishing through May and into June. So they got fishing before the water temperatures got up to where you couldn't do anything. With that thing not operating, and even if it is operating [the fishway at the Veazie Dam, which is at its most efficient at low water], they can't find the entrance. Those early fish that you'd like to have get upriver, the old genetic stock that we'd like to have, are either lost or there's damn little left of them.

We're taking fish to the hatchery now . . . all through the summer. So the bulk of the fish we're taking to the hatchery to spawn, to produce smolts or fingerlings to put back in this river, are fish that are taken during the summer run. So you expect to get later fish, and that really seems to be what's happening. During July we'll have a hell of a lot of fish come in the river in contrast to May.

Somewhere something in the formation of the Salmon Commission spells out their duties. They may have been reinterpreted since then. The purpose of restoring a run is to provide a recreational sport fishery. Providing for the survival of a species that may become extinct is all well and good, but if that's the case, that's a social program. This basically right now is a fisheries, fishing program. This is the one thing that's plaguing the whole fisheries and wildlife program in the state—the programs are supported by revenues from hunters and fishermen. Recently the budget of Inland Fisheries went from dedicated revenue to general fund. But there's a provision that the department must receive as a minimum the license fees. They have to do that to qualify for federal aid. Dingell-Johnson and this type of thing. But since they've been on the general fund they have not received any money from the general fund above the license revenue, except for a very minimum amount for search-and-rescue and two or three things of this kind. The study that was just done demonstrates that these resources in Maine are worth, what, $675 million, and this is accepted. The legislature commissioned this study. It's more than all the money paid for all the potatoes in Maine, by far. Many, many times that paid for all the lobsters and all the commercial fishes. It's seven or eight times more than that paid for all the dairy products. And yet the legislature has not seen fit up to this point to take any additional money from the general fund to help support the program. So, still, it is you and I and the others paying license fees. This makes it pretty difficult when it comes to talking about endangered species.

Jumping Salmon

From the collection of Bradley Stevens

Claude Westfall

Claude is a retired University of Maine professor. After military service and completing his undergraduate work at West Virginia University, he moved to Maine in 1952. He currently resides in Orono and is active in the Penobscot River Coalition to oppose the building of a dam at Basin Mills. He spends winters in Texas and, during the Atlantic salmon season, he can usually be found angling on the river at Veazie.

I first arrived in Maine in 1952 as a graduate student at the University of Maine. The Bangor Salmon Pool was one of the first local places I heard a lot about. I can recall, in those early days, going down to the Bangor Pool to fish for Atlantic salmon, knowing very little about salmon. I'm sure, looking back, that I didn't have the proper gear—I am not even sure if I had backing on my reel.

Those were difficult times. I don't recall how many salmon we had in the river in the early '50s, but the decline of the Atlantic salmon was quite evident. Salmon anglers on the river were few compared to today. Those who were well known include Guy Carroll, Adolph Fisher, Bill Walker, Frank Clarke, Arthur York, Ralph Getchell, Walter Coburn and Walter Dickson.

I was not successful in those early years and the salmon run continued to decline. There was a period during the late '50s when very few people were fishing salmon, and I gave it up. Poor fish passage and pollution from towns and mills up and down the river were big problems. Dissolved oxygen was so low below the Bangor Salmon Pool that some claimed this prevented salmon from coming upriver. The implementation of a water pollution cleanup bill passed by the Maine legislature in 1969 is responsible for the improved water quality we have today.

It was not until the late '70s that salmon began to make their dramatic return to the river. One of the most significant changes occurred when the Bangor Dam breached in 1977. This allowed free passage of all species of fish. A major shift in angling from the famous Bangor Salmon Pool to Veazie occurred and I began to take up Atlantic salmon fishing again.

I started fishing in Veazie at the Wringer Pool. Only a few people knew very much about this pool. The only boats I recall were those of Gayland Hachey and Buddy Lane, who worked for Bangor Hydro. Most people fished from shore and I remember Jim Walsh was one of the regulars fishing the Wringer; he and I would meet every morning. We fished this pool religiously, day in, day out, seven days a week. My first salmon was caught at the Wringer, a fresh 14-pound hen fish. After coming to Maine from West Virginia I confined my fishing primarily to the lakes in hopes of catching larger fish. Like many anglers, I found that after catching my first salmon my lake fishing was pretty much put on the back burner.

The fly I used to catch that first fish did not draw much attention. I went down the next day and took my second salmon, and was fortunate enough to catch salmon on each of the next four or five days. Anglers began to inquire about the fly pattern. I showed it to them and they asked, "What do you call it?" I responded, "I don't have a name for it. What's the name of the pool we're fishing?" Bob Ent, one of the few people who at that time knew the name of the pool, told me it was called the Wringer. This fly became the Wringer also, and I began giving them away. There were days that I did not have a fly to fish with myself. I would go home in the evening dead-tired after fishing all day, and tie up a few flies for the next day, and end up giving them all away. In the beginning I spelled the name R-I-N-G-E-R, since I did not realize the pool's history. It wasn't until some time later that I was told it should be spelled W-R-I-N-G-E-R, because there was an old washing machine wringer up on the shore that the pool was named for. I'm pleased that the Wringer pattern has become a fly that is used quite commonly on the river.

I fished almost entirely at the Wringer Pool in the early '80s. When the crowds began to show up, I decided to use my canoe, a 20-foot Grumman. I kept it on the shore just above the Wringer and when too many anglers showed up to fish the pool, I'd go up, launch my canoe and slide across the river to Dickson Pool, which would often yield a salmon. For the past few years now I've fished almost entirely from a 22-foot Grand Lake canoe.

If you go back to 1978, '79, '80, there were only a few people fishing the river at Veazie. As word spread about the angling opportunity, individuals from all over the state and the country began to make their presence known. The lines to rotate through the pools began to get a little bit longer, and it is not uncommon to go to the Wringer or Eddington pools today and find 17 or 18 rods in the rack. Conditions like this allow the angler to fish a pool only two or three times in a day. Long waiting periods to fish a pool have resulted in more anglers relying on a boat. It is an awesome sight to see the shoreline above the Eddington Salmon Club with as many as 40 or 45 boats on the shore. There have been as many as 35 boats attempting to fish a limited amount of water at one time, and some restrictions or controls are needed.

There is a lot of conflicting feeling between people fishing from boats and individuals fishing from shore. I hope people have learned that the water has to be shared. The pools can be fished by those who choose to use a boat and they can be fished by those who choose to fish from shore. It's been my observation that the boat fishermen are by and large pretty courteous and give more than those fishing from shore. Most boat fishermen are very careful to allow fishermen to rotate through a pool and not interfere with their casting. I think people have to realize that we have a limited amount of water, the fishing pressure has increased over the years—it isn't 1980 anymore—and we just have to share it.

Since the dramatic return of the Atlantic salmon in the late '70s and early '80s, I have been fortunate to take and release a number of salmon. This was possible because I worked at the University of Maine and was able to spend every day during the summer on the river. If you are out there a lot with line in the water, you are probably going to catch a fish or two. The other thing that has improved my success is being able to fish from a boat. This has allowed me to fish from the Veazie Dam downstream to an excellent stretch of water below the pipeline that is fished very little. I have had occasion to take a number of people out fishing, and knowing the pools is helpful. You know where all the lies are, where to drop anchor and where to put the fly.

I can remember when I started to seriously take up salmon fishing, I was like many fishermen—weighted down with vest pockets bulging with all kinds of flies. If I opened a fly box, no two flies seemed to be the same. People kid me today because when I open a fly box, all that will be in it, for example, are Wringers in different sizes. I believe in sticking to a few good patterns, in different sizes, that you have confidence in.

I have six fly patterns that I originated for the Penobscot: the Wringer, CZ Special, Foxfire, Pink Onyx, Penobscot Devil and the Green and Mean. I fish exclusively with these flies, for several reasons. I have been successful, and most fishermen will tell you it is a lot more fun to take a salmon on a fly you have tied yourself. I do not change flies frequently. Bob Enos, an old Gloucester fisherman and dedicated salmon angler, has fished with me for years and I always kid him since he leaves my Grand Laker full of little bits of leader, with his changing flies all the time. Bob has asked me about this and I tell him I need a good reason to change flies. If I raise a salmon that shows no interest in the fly, or I move to different water, I will change patterns or the size of the fly. If you are in a pool, know the water, know where the lies are, and if there is a taking salmon, you will probably find that any good pattern will take the fish. It amuses me, at the beginning of each new season, to hear anglers inquiring about some hot new fly to use.

Several events that remain very vivid include the day I fished with Dave Thompson in the Beach Pool. It was early evening and the water was very low, the rocks were showing and it was very treacherous to get the boat into the pool. After anchoring the boat Dave started casting, and I said, "Dave, put the fly right there." After he cast again I said, "No, it's got to be about 10 feet farther, that's where the salmon will be." Upon casting the fly about 10 feet farther, Dave hooked a salmon of about 12 pounds and we were off and running. The fish finally made a long run and left the pool; it was a circus to pull the anchor and get out of there. We got caught in the current and came broadside to a rock. Dave went head over heels and his rod came apart. With no strain on the fish his line went limp, and Dave said, "I lost my fish."

"Not necessarily," I told him, "I think your salmon is still on." I started hand-lining and could feel the fish. "Get your rod back together." Dave put his rod back together and I said, "O.K., now lift up." I finally got off the rock, and we bounced off another rock or two. Again, his rod came apart. We finally got ashore and Dave managed to land his salmon.

A well-hooked salmon will normally be landed even on those occasions when the line goes slack. When fishing alone from a boat, play your fish for a while, allow the line to go slack and most salmon will settle down. You can then start your motor, pull anchor, go ashore, lift up on the rod and the fun will begin again.

On another occasion I was sitting at the Wringer, waiting for my turn to rotate through the pool, and a gentleman came down, sat beside me and said, "I do not expect to catch a salmon." And I replied, "Well, you will not catch a salmon."

He asked, "What do you mean?" I answered, "If you ask people why it is that some individuals are successful and some are not, various reasons will be expressed. I happen to practice what I call my three C's. The three C's of Atlantic salmon fishing are: Covering the water. You watch some people go through a pool and cast one length of line, presenting the fly in the same manner all the time, not really changing the line speed and covering the water thoroughly. The second C is Consistency. You cannot fish for an hour and always expect to hook a salmon, although some people are fortunate to do that. You have to stay with it and not give up. The third C is Confidence."

"Confidence?" he asked. "Yes, this is the most important factor in success, in my opinion. What kind of work are you engaged in?" I don't recall what he did, but I asked, "Do you have confidence in what you do?" Of course he did, and I said, "I do not know any individual who is engaged in any kind of work who will do well if they do not have confidence in their performance. When I cast a fly, I always expect to catch a salmon."

He said, "You do?" and I replied again, "Yes sir. Every time I cast a fly, I expect to catch a fish. If you have confidence and are on top of it, observant, and covering the water, then your chances of hooking a salmon will be improved." I have always relayed this to people who ask, since these three ingredients are the main things I try to practice to catch one of the most unpredictable species of fish that I am familiar with.

The river is always undergoing subtle or drastic changes brought about by nature or man. A significant change occurred in 1983, when the Bangor Water District covered the pipeline [the water supply line for the greater Bangor area], crossing the river at Veazie, with 500- to 1000-pound stones. This essentially created a dam that ponded the water and changed flows at many of the pools. The Wringer Pool almost ceased to exist due to the reduced water flow. Navigation was hazardous and almost impossible except at high water. I filed a complaint with the Department of Environmental Protection and the DEP finally required the Bangor Water District to remove rocks to form a well-defined channel on the Veazie side of the river. I can recall, as large rocks were being removed, seeing a complete change in the flow of the water along the full length of the Wringer Pool. This saved the Wringer, but it is nowhere near what it was before the riprap work took place, in my opinion. The only gain was that it created an area along the pipeline for fishing that we did not have before. But we lost far more than we gained.

In 1987 the spring runoff was higher than normal. A great amount of material washed away from around the powerhouse at the Veazie Dam. Beach and B pools were altered and filled in. The finger bars on the Eddington shore below the Eddington Pool used to be well-defined and have eroded away. Flows have changed at Dickson. Without interference from man we will always have changes like these occurring.

The Penobscot presently has five dams in an approximately 50-mile stretch of water. It is discouraging after all the work, money and effort that has been put into the restoration program to find ourselves now faced with Bangor Hydro's plan to build another dam, at Basin Mills, and a new powerhouse—C-Station—at Veazie. The angling opportunity is limited, and C-Station will have a very detrimental effect on the water below Veazie Dam. The Eddington Pool in all probability will not exist and water flows will change drastically. Once you have something taking place on the river and man interferes, by building a new dam or fishway, it's just not the same and it will never be the same. The entire ecosystem is affected, and conditions are seldom what are predicted. They are always a little bit different.

Several of my questions and concerns for the river include: The protection of the river, at least from the Veazie Dam down. We cannot afford to go backward by losing additional angling and spawning habitat. One of our biggest problems is downstream smolt migration. Mortality is too high and if this is not corrected soon, restoration of the salmon will not occur. Fishermen have been releasing more salmon to spawn upriver. The stocking of smolts remains high, but the total number of returning salmon is not going up as should be expected. What is happening?

I think our fee structure needs to be revamped. Residents pay $10 and nonresidents $30 for a salmon license. The opportunity to fish the entire season for these low fees is ridiculous. Things have changed. If we are really interested and committed to restoration, we as salmon anglers should be willing to pay more for the privilege. We are losing our natural resources like clean air and water, wetlands, soil, forests, grasslands, fisheries. It is now time to be willing to pay a little more for it.

The progress that has been made in the restoration of salmon on the Penobscot has been made possible through the efforts of many agencies, organizations and dedicated individuals. The Maine Council of the ASF has done a great job, and the present leadership has been excellent. I look back to when the council was first formed in 1983. There were approximately 10 of us, as I recall, whose names were suggested by Bud Leavitt. We met in Bangor with Wilford Carter, the Executive Director of the ASF, to form the council, and Walter Dickson was elected as our first president. The organization has grown in stature and it was nice to play a small role in its organization.

Unfortunately, biological diversity, a loss of plant and animal species, has become one of our most pressing global ecological problems. The future may not appear to be very bright at times, but we must continue to oppose those projects that threaten our quality of life. We can now look and see our feet in the Penobscot when standing in the water. Let's hope we can continue to see conditions improve—that's the big challenge we face and my primary concern for the future.

Action at the Pipeline

Walt Dickson

Past president of the Hildreth Network (Channel 5—WABI—WAGM) in Bangor, Walt helped start Maine's first television station, Channel 5, in 1953. He caught the Presidential salmon that same year, presenting it to Dwight Eisenhower. A recipient of the Roll of Honor Award from the Atlantic Salmon Federation, he was the first president of the Maine Council of ASF.

I think I first got interested in salmon fishing back in the '30s, when I was just out of high school. I was fishing with an old-timer down in Hancock and we were fishing Tunk Lake for landlocked salmon. About noontime he said, "Let's go down to the Narraguagus River. They have Atlantic salmon in that river." I suppose I had heard of the Atlantic salmon, but I didn't know much about it. So we went down and we stopped at what is now called Academy Pool. This was in May, I think it was Memorial Day, and the salmon were jumping and rolling in Academy Pool on the Narraguagus. All over the place. We drove up and down the river and didn't see more than one person fishing.

I decided, *Gee, I've got to try this.* And I can remember him saying, "They'll take something red, they'll take anything that's red." It's been so long that I don't remember, but a year or two later—it might even have been that year—I started fishing the Narraguagus. In those days there were a lot of big fish, and a lot of fish compared to the number of fishermen. There'd be only four or five fishermen on the river, and always the same old-timers—Harry Smith and some of those fellas.

I finally caught a salmon, and in the '40s I really started catching fish. In 1947 Horace Bond, who was really instrumental in bringing this river back, asked me to serve on the Atlantic Salmon Advisory Commission Governor Hildreth was forming. Horace was chairman, and I think there were about seven or eight of us, in those days, that served on it.

Of course, Horace's interest was in all Maine rivers, but mainly he had a great belief in bringing back the Penobscot. Not many of us believed it was possible but he never let go of that. It was going fast then. In those years I never saw anyone fishing from the shore at the Bangor Pool, but they'd be out there in the boats rowing and catching salmon. Then it kept getting worse and worse and worse and by the time the '50s came the salmon fishing was gone.

In May of 1953 I caught the Presidential salmon, from the shore, and the fishing was just about over. The records show it had just about petered out. Then there was a period of twenty-some years with no salmon at all.

In the late '40s and '50s the river was just like a sewer. It wouldn't be too bad with the tide going out, but when the tide came in the sewage from the city and the hospital would float right back up into the Bangor salmon pools. It was really a mess in there. And it got worse.

And then of course the salmon came back. The salmon of the Penobscot came back in the '70s. It was just unbelievable when those fish started to come back. And once the towns started cleaning up the sewers and the mills started cleaning up, the water in the river was noticeably cleaner. Then the late '70s, '77 I think or '78, the breach in the Bangor Dam appeared and that changed everything on the Penobscot. That's when fishing started again. The fish could get up to the Veazie Dam and beyond.

We stocked fish for years. I don't think they ever stopped stocking this river. I think they were always putting some in. But they started stocking heavily again in the '70s, as I recall. Of course, they are stocking it heavily now. I think the agreement with the Feds was that we'd stock 600,000 smolts in this river once it was declared a restoration river, and the US Fish & Wildlife Service is still doing that, approximately 600,000 smolts each year.

The Maine Council of the Atlantic Salmon Federation has done a lot of good work and they're very active. We set up the Maine Council in the fall of 1983 and I was elected the first president. In a weak moment I accepted that, and I served for two years. Then Malcolm Coulter took it over for two years, and now Bill Townsend is president. We helped get some new regulations passed, and some still need to be changed. We've been able to get the limit on the Penobscot [and now, as of 1992, statewide] down to one fish just to protect the river. We've got registration of fish, which we didn't have. But we are after more revisions of laws on Atlantic salmon. Some areas are very confusing, especially on the netting in the mouths of the rivers and off the coast. Hopefully we'll get that squared away. I think the registration is going to help determine just where these fish are being caught, who's catching them and the size and so forth. We're not against all the dams but we are very active in making sure that these dams, new dams or present ones, have proper fishways that are maintained properly. That hasn't always been the case.

About three years ago the Atlantic Salmon Federation began to promote a catch-and-release program. When I was fishing in the '30s, '40s, '50s and '60s, and even into the '70s, when we caught a salmon we always used a gaff, never a net, and it was kind of a brutal job. I never saw a net on the Narraguagus or any of the other Maine rivers until the '60s or '70s. Then a few nets appeared. I remember some of us being amused when some chap from Massachusetts showed up on the Narraguagus River with a net. We wondered what he was going to do with it and how he was going to catch a salmon with a net.

The Atlantic Salmon Federation started pushing catch-and-release and in the last three years it has caught on, I think, on all Maine rivers and especially the Penobscot. Now about 50 percent of the salmon that are caught are released and that helps the restoration program and changes the whole picture of catching an Atlantic salmon from just killing it to letting it go up the river to spawn and perhaps be caught again. Now fishermen are even hand-tailing salmon, and learning to keep them right in the water, pulling the hook out or cutting the leader—that's saving a lot of fish.

The Maine Council has been active in other ways. About five or six years ago Swift River Company decided they wanted to rebuild the Bangor Dam, and if they had done that then the fishing from the Veazie Dam to the Bangor Dam would have been destroyed and all this great fishing area we have here now, about four miles of river, would have been lost. The clubs got together and we formed a group called Friends of the River. The purpose was to defeat the building of that dam and I think we have been successful over the past five years. The group has raised $70,000, we hired a law firm to represent us, we attended the hearings, we spent a lot of time going around the state explaining why a dam would not be beneficial and would be damaging to the Atlantic salmon restoration program. Hopefully we have won that case. We won the case as far as the Department of Environmental Protection was concerned. The Swift River Company has now taken it to court and we expect to get a decision sometime in January. Hopefully it will be in favor of the Atlantic salmon. [The court denied Swift River's application.]

The Friends of the River is still active. It was formed specifically to stop the rebuilding of that dam, but we probably will continue to be active and will take a part in removing what is left of that old dam so that we'll have some new pools and a free-running river from Veazie right down to the sea.

The Swift River people claimed that they were going to come up with some new modern fishways, which they never produced at the hearings. We learned that even the best fishways now available will allow only about 80 percent of the salmon through. Each fishway loses up to 20 percent of the fish trying to pass. Swift River never produced anything that gave us an idea of what type of fishway they were going to have or what they were going to do, but it was proven, I think, beyond a doubt that every time you put a fishway in a dam you lose anywhere from 10 to 20 percent of your fish. I think that had a great deal to do with convincing the DEP people that Bangor Dam would not be beneficial in the restoration of Atlantic salmon. I think that is why we got a positive vote from them.

The people from Swift River were saying they could build salmon lies and telling us where salmon would rest, but you know that would have been almost impossible under the conditions if they had put that dam in because the water would have been too deep. You wouldn't have had the proper flow of current and you've got to have the right depth and flow of current to find the salmon lies.

We would have had a lake from the Bangor Dam to the Veazie Dam. When I moved on the river

15 years ago, the Bangor Dam was in fairly good shape and it was simply a lake out front. I never had any idea when I first moved here that there would be salmon pools on this bank of the Penobscot, because the water was almost up into the trees.

When that breach came in the dam, as I looked out here I began to see places that looked like they might make pretty good salmon pools. That's of course when we first learned the salmon were here.

I watched one particular area as the river dropped, where a brook comes in, and it appeared to me that salmon should lie right in there. I kept trying it and one afternoon bang! a salmon was on. There have been salmon, I guess, in that pool ever since. [It has since become known as Dickson Pool.]

I had relatives right on the Miramichi, at the mouth of the Caines River. I fished there every year for over 25 years. I thought about how wonderful it must be to be on a river where you can walk down over the bank and catch a salmon. Then all of a sudden I found myself living on a river with Atlantic salmon, and some of the best salmon pools in North America. I know now how it is to live on one.

I remember one day I was fishing up on the Eddington shore and this chap came along who I later learned was Leroy Hutchings. He had a handful of little yellow slips of paper. He said, "I'm taking names of people who would be interested in forming an Eddington Salmon Club. Would you sign your name here?" And I thought, *Why the hell is this guy talking about forming a salmon club? We've got two salmon clubs here. He'll never be able to form a salmon club in Eddington.* But I said I'd sign my name to it anyway, never dreaming of what did develop. I have to give credit to Leroy. That club of course now has 400 members, and is the largest salmon club in the world. We're debt-free and built a new clubhouse in 1990. I think more than 20 percent of our membership is from out of state, and they have been among the best contributors to the club and to Friends of the River, and to helping keep the Penobscot restoration going. You never can tell when you sign a piece of paper what will happen. It turned out to be a really great club and a really great group of people.

There are always going to be threats to the river. There's a new proposal now to add a new power station, C-Station, to the Eddington side of the Veazie dam. The station could ruin the salmon pool on the Eddington shore. The Maine Council is involved in that now, trying to determine what will happen and what can be done to prevent that.

The proposed dam at Basin Mills, about three miles upstream of Veazie, would be another obstacle to the passage of Atlantic salmon. The Maine Council and the Friends and the Penobscot Coalition are active in determining what can be done to prevent it or to take proper steps to get assurance that the fish can pass without any losses. If it is built it will prevent many fish from moving upriver—that's just one more dam, one more thing in the way. It's going to need a lot of help, a lot of people, a lot of money and we're going to have to be ever watchful if we are going to restore this river.

You can never rest. Just to maintain what we have is a big job and it takes a lot of people and a lot of help. I thank the Atlantic Salmon Commission staff—I think we've got a great group over there. But they're short-handed. They need more funds, they need more help. The clubs and the people should be going down to the legislature in Augusta and helping the Atlantic Salmon Commission get the funds they need to restore these rivers.

I think we need a lot of research and work on regulations. We've made progress with the one-fish limit. The other thing is fall fishing. We don't have a fall run on the Penobscot and I don't think they should keep the river open after about August 31. All you are doing up to October 15 is catching dark fish that have been in the river some time. I know it wouldn't be very popular but I think it would save a lot of fish and help the restoration program.

Bob Ent

Known most for his Atlantic salmon flies in the classic style, Bob was also the originator of the Pink Ent, a pattern still popular, and effective, on the Penobscot. He was also a soft touch for anyone who was short of flies on the river. He passed away shortly after this interview.

When you first caught a fish up here, everybody would be standing around watching that fish being caught, because it was so spectacular to see a fish come back to this river. We only caught six or seven in a season, and that was doing good. Then it got to be where maybe they would catch two or three in a week, and that was still a big thing. People would really gather when they heard there was a salmon run on the river—they would come from everywhere. And there weren't any fish, of course, above the Bangor Dam. It was all right there around the [Penobscot] salmon club, below the Bangor Dam.

Now the salmon can get back up river, and it is common to see a fish caught. Nobody even thinks anything of it, until you get a bad year like we had this year [1987]. A lot of people didn't catch any fish this year, and now they respect the fish a little bit more. Maybe next year they'll catch all kinds, and then they'll treat them kind of like trash as they catch them. Just turn them loose, like there was nothing to them, without recognizing how important they are. That is something I have noticed about the restoration of the river. Although it did a lot to bring a lot of people together, they don't respect the salmon as much as they did when it first started. If you caught one, boy you were really something;

there was something very special to it. Now you can catch one and just turn around and say, "Oh, that's only an 8-pounder," and turn it loose like it's nothing.

When someone had a fish on, everybody seemed to know it, and they all went down to the shore to see what was going on. Now it's not so unusual. And we have a different group of people on the river. We used to have only a few who really knew how to fish for salmon. Now, it seems like we have more instant experts than we ever did before. That doesn't mean anything, but there are a lot of them now. Some of them have a lot yet to learn about salmon fishing.

One of the things that has happened, is that it has gone completely back-asswards. When we first started catching salmon [in the 1970s], they killed every salmon they caught. You were allowed two a day, and usually all of the fish were killed. Nobody turned fish loose. There were a few that I can remember, and not on this particular river, but Dr. Babson [on the Machias] would always turn fish loose way, way back when it wasn't popular to do so. They thought he was crazy. Well, he knew what he was doing.

After a while, some people started turning a few fish loose over here on this river, and it started catching on a little bit because they were catching too many. Then the regulations changed, and we could only kill one fish. It was surprising how many people released their first fish anyway, and it was tough for them to do that.

A lot of people gave up salmon fishing because of the new laws, but many of them have come back to it. After they released one fish, they felt so much better after they did it that they started doing it all of the time.

The only requirement I would like to see, especially on the Penobscot, is after you keep a fish, you have to use barbless hooks. But it would probably have to be voluntary because I don't think they can make it a law. It is a good practice, and you're just fishing for the fun of it anyway. The people who are fishing should say, "Well, I'm not going to be able to kill anymore anyway, so I might as well bend my barb down and if I get one, all I've got to do is turn him loose." I've seen fishermen having a hard time getting fish off barbed hooks. If you're in a boat you don't even have to bring the fish into the boat with you. You can just run your hand down the line until you touch that hook, and give a little twist, and it's gone.

Having more salmon in the river is of course a boon to a lot of the fly shops. There are a lot of people tying flies commercially now. I don't tie for that reason at all. My primary reason for tying flies is because I like to do it, and I've been doing it for—I've got to stop and think this over—about 40 years. I started with trout flies, and them moved into salmon flies.

I think you always come up with a few little differences in patterns, but if we'd research all the books that there are, probably everything that is new has been tied before. It just might be called by a different name, but it is probably there. You called it a Black and Tan, and it used to be called a Tan and Black.

Everybody still likes to have their own little thing, so it is intriguing that way. You tie one up and you think, *This is going to be it.* You go out and you catch a fish, and you think you've really got it, then you go the next four trips and you don't get anything, so you go back and tie a different one.

Having the Penobscot get better and better has probably kept me interested in working on my fly patterns because it is handy for me to check it. I would still do it, though, because I fished for salmon in Canada before the Penobscot came back. Oh, we had the Machias and the Narraguagus, but they would only hold so many people and it was kind of selective as to who fished where and what and when.

The river coming back was a wonderful thing to see. When I came here there were very few fish, and now to go over to the river and see maybe 20 fish per day, or maybe sometimes more, is just unbelievable. In fact, I didn't even fish the river when I first came here because it was so bad. And I think most of the people you see over there are realizing it too. When the salmon began to come back, you started seeing more canoers, or families coming down to the edge of the river for a picnic, when they wouldn't have gone near it before because of the pollution. I think it is a wonderful thing to see.

I think one of the great things we also witnessed is that hole in the Bangor Dam opening up, and then getting bigger over the years. One of the next great things would be to get the dam out of there.

We have a hell of a battle ahead of us, because there are going to be improvements on the dams that are already there, and another effort to rebuild that one wouldn't surprise me a bit. Everything is under a constant threat.

I was talking yesterday with someone about how many people have gotten to know each other through the salmon clubs right in Bangor. I would probably never have met a lot of people if it hadn't been for the clubs bringing us together. It happens through the clubs and on the river, but most of all through the club machine. You might see somebody on the river and ask who it is, but when you go to the club meetings and you hear somebody or you talk to them, that's when you get to know them. I think the clubs are what pulled the people together and now we have three right here in Bangor. When they started the Veazie Club there was a little bit of ill feeling in the Penobscot Club. It used to be that the people on this side of the river didn't like the people on that side of the river. They got over that and now, with the Eddington Salmon Club, we have three.

With anything that has to do with that river, they are all together. There will be little spits and spats, but that's just politics. When it comes to a real show-down they hold together. That's when you need it, because we'll get more done with the clubs than we will without them. Without these clubs I don't think they would have stopped Swift River [from rebuilding Bangor Dam] at all.

It's great to have salmon fishing right here in the States, and not to have to go to Canada. Salmon fishing in Canada, as you well know, costs a minimum of $30 per day for every day that you fish, just to get permit water. I guess you don't have to have a permit to fish the Margaree, but I understand that next year you will have to have a guide, so one way or the other they are going to get that $30 to $40 per day from you. That is something you don't have here in Maine. Once you buy your license you're ready to go. No charge whatsoever. When I say $30 a day in Canada, that is a minimum charge. You can go as high as $400 to $500 per day, depending on what you want to pay. Yes, this is a wonderful thing that they have got in their backyard.

It speaks well of the vision of those who saw the potential for this river early on, and you've got your modern-day people that have the same vision now. You've got Tom Hennessey, Bud Leavitt, Walter Dickson, all of them. They can see ahead and see what the river means. Christ, there are so many of them I can't even try to name them. They are all following in the same footsteps the others started. All they are doing is making it come to life now, no matter how they do it. Hennessey does it with his paint brush, and raises money for the clubs. Leavitt does it with his writing, Al Meister does it with his expertise on salmon restoration. Then you've got the individual fisherman—you say something to him about something happening to this river and he is right up in arms.

It's also meant a lot to me personally. Where else could I go when I want to go fishing for half an hour? Just take my rod and tackle, get in my vehicle and drive over there and maybe catch a fish, think nothing of it. Come back in two hours and go again, and then again. Maybe I'll go fishing a dozen times and never catch anything, but it is still there if I want to go.

The Eddington Salmon Club

Jane Cleaves

Jane Cleaves grew up on the south shore of Long Island, New York, and spent many hours fishing and boating on Great South Bay with her family. In the early 1970s, she moved to Maine while attending the University of New Hampshire. While working at L. L. Bean, she became involved with a number of sporting and conservation groups, among them Ducks Unlimited, Trout Unlimited, the Ruffed Grouse Society and the Maine Council of the Atlantic Salmon Federation. Now living in Bowdoinham, working for the ASF from her kitchen, Jane lives with her chocolate lab, Amherst.

I always fished growing up, and when I first started working at L. L. Bean—in 1980—I'd been trout fishing in New Hampshire as a college student. I was hired at Bean's as a camping and fishing clerk. There was a wonderful guy working there at the time named John Mayo. John lived in Yarmouth, and he was a good friend of Bud Leavitt's. John fished the Narraguagus, the Dennys, the Machias and East Machias—he fished a lot Downeast. He caught an 18-pound fish on the Narraguagus that Bud wrote about in the *Bangor Daily News*.

We'd always talk about fishing or hunting in the store, but whenever we'd talk about fishing, John would talk about salmon fishing. Just listening to him got me hooked. We came up here to the Penobscot a couple of times—this was in 1983 or so—and at one point I got hooked up with Arthur Taylor. That got me involved with the Maine Council-ASF and the folks up here.

Being involved with the Maine Council was the natural lead-in to the Federation, because the Maine Council and the Federation work so well together. Through Maine Council functions I met Bill Taylor, back in '88. He had just been hired by ASF and was doing the regional council work that I do now. He covered Maine, New Brunswick and Nova Scotia, all of Atlantic Canada. Danny Bird was later hired as the first regional coordinator; Danny covers New Brunswick and Nova Scotia, and I'm Maine and New England. There are two other coordinators for Quebec and Newfoundland/Labrador.

I was involved as a volunteer, organizing banquets for the Maine Council, and over a year ago the Federation was exploring the possibility of having a Portland banquet. My name got on the list for the committee, I became co-chairman with John Swan, and last December I was approached by ASF to work full-time.

My primary functions are working with the Maine Council and the affiliate organizations, public education and awareness of ASF and its programs, membership, banquets and fundraising. I'm doing everything that ASF mandates on the grass-roots level, working with the volunteers and helping them be as effective as they can by staying on top of the issues in their region.

The New England Council is in the early stages of development. There was a New England Council at one time, but the organization may not have been structured as well as it could have been. Once a council is established, it takes about a year to get it fully organized and up and running. The Maine Council took about a year and a half to get going and be organized and effective. We have 19 affiliates in Maine right now, the most recent being the Downeast Salmon Association.

Two delegates represent a club, and I represented the Merrymeeting Chapter of Trout Unlimited. The responsibilities for the delegate are to attend council meetings, which are every other month, and report back to the club about Maine Council activities. In addition, we in turn would bring to the Maine Council concerns and goings-on of the club. It's a communications link back and forth. I was nominated to be a director of the Maine Council two years ago, served as a director for two years, and was elected vice president of the council last fall, which I had to give up when I became ASF staff.

I am the only staff person in Maine and New England. Our staff in St. Andrews is at approximately

40 people, and we also have a research laboratory in St. Andrews, which is our Salmon Genetics Research Program. They too have a staff there that runs the hatchery and research center. It's quite a big complex in St. Andrews.

The Maine Council has been in existence since 1986. ASF came together from two of North America's biggest salmon organizations—Atlantic Salmon Association and International Atlantic Salmon Foundation. The idea of regional councils in all of the provinces and New England quickly developed, and ASF has six regional councils. This past weekend I was working with the three other regional coordinators and representatives from the regional councils of New Brunswick and Quebec. We had meetings over the past four days, formally and informally, and it was good for the Quebec and New Brunswick councils to get a feel for what the other councils are doing. Every council has very different problems and challenges. The Quebec council is well organized—they have a lot of members, they have good fund-raising ability, they're very highly developed, and while they're all volunteers, they're at a point where they have a paid director. New Brunswick is much the same. I'm talking about something like 80 clubs.

In Maine we have 19 affiliates, but Maine has been very effective. In the past four years Maine has had two affiliate-of-the-year awards from ASF: The Friends of the Penobscot River for defeating the Swift River Company, and the Penobscot River Coalition. Maine has been very powerful and effective.

Before ASF, Maine was essentially on its own. There were members of ASA and IASF in Maine, but there was no linkage between the clubs and the organization. The Veazie and Eddington salmon clubs weren't here. The Penobscot Club was, and some of the TU chapters were, but they had no linkage, and that's what the regional affiliate program does. It's a very strong, effective program.

I think the restoration is going to work. I get charged up every time I go up to New Brunswick and see what those people are doing, and I think it's only a matter of time for us. We're heading in the right direction, we're just a little bit behind them. We have just as many dedicated and hardworking people as they do—but I think it's important we keep the restoration movement alive on the state and federal level. That's where I see we could run into problems. We can't lose the support from there, because that will kill it.

We need more lobbying efforts, and not just in Augusta. Ideally, we need someone lobbying in Washington. We do have some good news coming down from Newfoundland. They reduced their commercial quotas by 10 percent this year. That's not a heck of a lot, but it is a reduction. The quota will be about 600,000 metric tons, down from almost 700,000 last year. The other good news from the provincial government indicates that they will support a buyout next year in Newfoundland, so this could be the last year of commercial fishing in Newfoundland. There is also talk of volunteer buyouts in Labrador.

It's another piece of the puzzle. We had the Faroe buyout, now we have good news coming from Newfoundland, and hopefully we can get some agreements up in west Greenland with NASCO in Edinburgh in June. We have Bill Townsend there going to bat for us. All these things are very important for the salmon down here. They're all important for the success of the restoration.

Al Meister

Al was chief biologist with the Atlantic Sea-Run Salmon Commission from 1967 through 1988 (the year this interview was conducted); he joined the Commission in 1957. As chief biologist for more than 20 years he oversaw, and was partially responsible for, a tremendous improvement in the Penobscot's salmon population.

I started with the Atlantic Salmon Commission in 1955 as a Research Fellow at the University of Maine, sponsored by the Sport Fishing Institute in Washington, D.C. I then went on as a Research Assistant. In 1957 I joined the staff full-time as biologist, and in 1967 I became chief biologist, with complete charge of the program.

Everything takes time, and I'm a bit disappointed in the length of time that has been required to reach where we are today on the Penobscot. But, in hindsight, there has been steady progress. When I started in the mid-1950s my major activities were a survey of the headwaters and tributaries of the Penobscot River and at that time there were no salmon in the river, no salmon in the bay. The run had been lost.

We felt we had an opportunity to do something that had never been done before—to restore Atlantic salmon. Now, there are two things that must be kept in mind, that were very fortuitous: Both the state and federal governments were interested in the pollution abatement program. [The federal government] passed the Clean Water Act of 1968, although they didn't fund it. However, in 1969 Maine passed a bond issue of $50 million—which represented, at that time, an assessment of more than $50 for every man, woman and child in the state—for clean water, and that combined with the Federal Clean Water Act, which was eventually funded, permitted us to clean up the Penobscot River. In 1968 the Penobscot River had a pollution load that was equivalent to the combined domestic pollution of Rhode Island, Connecticut, Massachusetts, New Hampshire and Maine. With a population of less than a million here in Maine, the Penobscot alone received a pollution load that was equivalent to the sewage from about eight to 10 million people.

Although the towns themselves were dumping raw sewage, the majority of it was from the pulp and paper companies. Had they been located on one stretch of the Penobscot watershed they would probably have cleaned up their own mess many years ago in order to have clean water. But they were spread out enough that when the water reached a mill, it had cleared up enough so they could use that same water and repollute it. The river never had a chance to cleanse itself.

Along with that, the federal government passed the Anadromous Fish Conservation Act in 1965, although it was not funded until 1968. Here on the Penobscot we had, like a storekeeper, made an inventory. We knew what we had to work with, what we had to sell, and we were able to write a restoration plan that was submitted to the federal government—the Model River Plan for the Penobscot River. Acceptance of this plan provided the funds we needed and these monies were matched by both industry and municipalities to build fishways at the dams on the Penobscot.

The industries on the river, whether it was hydro or pulp and paper, met us halfway. They said, "You say you need two fishways on some dams because of the configurations? We won't build two, we will build you one. Pick the one you want, and when it's used then come talk to us again." We are now reaching that stage—those talks are under way. In places like West Enfield and Howland we are getting new fish passage and we will probably have another fishway at Veazie in the not-too-distant future.

I can't fault industry for not wanting to invest their stockholders' money without some sort of guarantee on a return. I think we have proven that we can get fish back. As you are aware, we have fish that are making it into the East Branch of the Penobscot, the Mattawamkeag, and the other headwater tributaries where pristine spawning and nursery areas exist. This can't do anything but help. A major portion of the run, up to now, has been of hatchery origin. There may be times in the future when it won't be primarily hatchery stock, for the wild stock is going to play a larger and larger part in the restoration.

As near as I can calculate—my figures have been checked by some of my colleagues here and in other parts of the world—historically we probably had a run of something in the neighborhood of 70,000 salmon in the Penobscot. Grilse were almost unheard of, so these 70,000 fish were large salmon, repeat spawners, many of them probably in the 20- to 30-pound class. The average fish in the old commercial fishing records for Penobscot Bay was 16 pounds, and I feel sure that those fish were the ones that went into the West Branch of the Penobscot. Now that area has been removed from the restoration list because of the existing landlocked salmon fishery, the large lakes and the difficulty in fish passage. The run in the Penobscot will probably be, in some years, 50- to 75-percent wild fish. There will be other years when the majority of the run will have to be hatchery fish. I say that based on present-day technology. We lack adequate engineering expertise and the technology to provide good, reasonably priced downstream passage for fish.

If you check the US Geological Survey water records you will find that there are occasional years when most of the fish moving downstream will be forced to go through the hydroelectric turbines. When that happens, you have a very severe mortality that will be reflected two years later in low returns. So in those years most of your run is apt to be those of the hatchery returnees that are released from the lower end of the river, where they escape the turbines. With better technology and downstream passage, we could eliminate a lot of that and perhaps level it out and increase the wild portion of the run to maybe 80 percent. But right now I am looking at a 50/50 mixture. In round numbers, that probably translates to a total run of somewhere between 6,000 and 12,000 fish. Hopefully, I am going to be right. It is better than we have done to date.

Projections for a coming year's run can be tricky. I had a reasonably good track record for a few years, but with the closure of the Greenland fishery in the late fall months, and the closure of the fall fishery in Newfoundland, I lost two of the indicators I used to determine what we were going to get back home. So instead of being able to gauge the returning run on the basis of a six-month lag in time between the fishery in Newfoundland and when the fish come home, I now have to operate on a 12-month lag, and when you look at the increase of sea mortality over a 12-month period it will sometimes throw you for a loop.

Without the input and support of a small group of Maine sportsmen, there would be no Atlantic salmon in the United States. For years the federal involvement was very small, confined strictly to raising a few fish in the hatchery for use in our restoration efforts in the small coastal rivers Downeast. When we went to the legislature for our operating budget and funds, we had to rely on the sportsmen to support us and *they did.* They have supported both the preservation of the fish as a species in the United States, and the conservation of the fish within the state. They, themselves, have fought for and supported the reduction of limits. When I first started there were two fish per day, and no limit on the season catch. If you had people who took 50 fish a year, they would be considered fish hogs today, but without their support, we wouldn't have a program, there wouldn't be a sport fishery.

The sportsmen changed their thinking and realized that they had to reduce their harvest as individuals, but they also had to work for and support the program if they wished to continue to be salmon fishermen. It has been overwhelming at times, the support that we have had, and we are down to one fish a day now, and five—and in some areas one—per season [now it's one fish per year, statewide].

I must admit that, personally, I was a little opposed to reducing the season from an April 1 start to May 1, because the fish taken from April 1 to May 1 were not needed when you look at it from a biological standpoint of spawning stocks and escapement and the rest of it, and there were an awful lot of people who couldn't wait to get their hands on their fishing rods, to put their waders on and go fishing. We removed a large number of days of fishing opportunity by reducing the season by that one month. The same thing could happen in the fall. I'm not sure we should permit fishing throughout the entire length of the river, but there is no reason we shouldn't have fall fishing on a catch-and-release or limited-take basis from the lower portion of the river. I want to preserve the fish, but if the individual likes to get out in the autumn coloration and fish on a clear frosty morning, he should be allowed to. And without the active angler on the river, it is much more difficult to enforce [regulations]. An angler on the pool will keep the poacher away. It is when there is no one there and our enforcement people are tied up in the fall with migratory waterfowl, big-game hunting and the rest of it that the poacher can have a field day.

The whole program, in reality, goes back to the sincere efforts of a very dedicated individual, Horace Bond, who was not a trained biologist, he was just an ardent salmon angler, and he felt that something was remiss when we did nothing to protect or promote the salmon resource. He worked very diligently for the resource and was the first chairman of the Commission and served faithfully for many years. He was one of several who were instrumental in getting the legislation passed that established the Commission. I think we owe a great deal, and the sportsmen of Maine and the United States owe a great deal, to Horace and to the many others like him who had an ardent love affair with *Salmo salar*.

Sportswriters like Bud Leavitt have provided us with excellent publicity over the years. Without their help, we would have been required to provide our own PR program. When we first started getting fish back to the Penobscot, we had no real laws to protect them at the Bangor Pool. There hadn't been any fish there for over a decade so there was no reason to have any laws. Kids will be kids, but some of the kids were 70 and it didn't take them long to find the fish in the colder water of the tidal pools and the mouths of the small tributaries. What they did to those fish would make an ardent fly fisherman cringe, but we needed the publicity and we received publicity, countrywide. Bud Leavitt called me and said, "We gotta do something about that Al." We had authority to enact emergency provisions to close it to protect the fish, and I can perhaps be held responsible, but I said, "No, let it go for the weekend," knowing full well that there would be a horror story to be told on Monday.

Poor Bud's phone, I guess, rang off the hook all weekend. Neither he nor his wife got a minute's sleep, with people complaining about what was going on at the Bangor Pool, but when the news story broke, around the world, Bud was able to write his column and tell the people on the phone that yes something was being done. We received both the publicity and the support we needed to promote the program and go from there, and that is how it really got started. Whether I should be chastised for letting them kill a dozen fish, I don't know, but because of that we've got several thousand fish today, where we had a few hundred then.

The existing fresh-fish market that reaches interstate and international commerce is being provided by the aquaculture industry raising Atlantic salmon. They are available 12 months out of the year, they are uniform in size and color, they bring top price. The wild stock is still being exploited commercially. They are being exported from Greenland almost exclusively to provide the smoked fish industry. The biggest market of course is in Europe, and to the best of my knowledge all the Greenland catch today is frozen, then exported to Scandinavian countries for smoking. None of it is sold fresh or as salmon steaks, it goes for smoking. I think there will continue to be a modest harvest of wild fish by these commercial fisheries. I don't think you will see them leave—they are too important to the native populations—but no way will they ever be able to supply the demand for fresh salmon that is being met by the aquaculture industry. The world catch of wild salmon is running in the range of 9,000 metric tons. The aquaculture industry will, before the end of this decade, exceed 90,000 tons. There is no way the wild stock could provide that . . . that is more than they have ever provided, even historically.

I can see a curtailing of commercial fishing as such continuing in the Maritime provinces and the Canadian mainland, in the UK and in some of the Scandinavian countries. Norway has now outlawed fishing in the high seas off her coast for her own people, and I think there will be a continued reduction in the number of licensed commercial fishermen, with one exception—that is Ireland. They don't know how many people are fishing there. But eventually that will come under control, and in the same way the North Sea fishery off the coast of England and the southern coast of Scotland will be reduced. It will come because they can't compete with the aquaculture industry.

My favorite Penobscot fishing story involves a man that you know quite well, of French extraction, and he was fishing from one of the old double-ended peapod rowing canoes that they always used on the Penobscot and the Bangor Pool. He had found one and was able to resurrect it—there are still two or three in existence, and some new ones have been made since then. But the seats had deteriorated with dry rot and it had become quite unstable. He was fishing in the Bangor Pool in the traditional fashion, rowing the double-ended canoe, with the rod trailing a fly behind, not really trolling, but letting the current work the fly, and a fish took. In his urgency to make sure that the fish was properly hooked, and to get headed for shore (they never fought the fish in the canoe, they always came ashore to fight the fish), the seats of course let go, and this salmon angler ended up in the bottom of the canoe, but the rod went out over the end of the canoe and was last seen being towed downriver by a very irate salmon, who jumped several times.

The man, fortunately, didn't upset, he got back and was able to get the oars and get himself ashore. There were several people sitting on the porch of the Penobscot Salmon Club watching the whole escapade, and when he came ashore, and looked up and saw them watching him, he reached down and picked up a piece of driftwood, which happened to be a 2x4 on the shore. He walked up and his first words were, "I don't want to hear one word about how a Frenchman fishes."

I must say that I have enjoyed the people that I have worked with, enjoyed the sportsmen, and the support the programs have received, and I think that in the not-too-distant future, that I am going to back out and participate in the sport fishery that I helped establish.

The Ledges at Eddington
From the collection of Dave Meck

Dorothy Douglass

Born in Kansas City, Missouri, Dorothy was raised in Coffeyville, Kansas. She was a computer programmer for American Airlines, and ran an antiques business in California before retiring to Bucksport, Maine, with her husband, Ed. Besides being a tier of fine Atlantic salmon flies in the classic style, she is currently treasurer of the Maine Council of the Atlantic Salmon Federation.

I've been tying classic salmon flies for about seven years. Twenty years ago or so I took a class and tied some trout flies, but I never did it again. When I started tying again, I started with classic salmon flies. One day, after my husband and I retired to Maine, I happened to find a book on classic salmon flies and I thought, "I've just got to learn to tie those."

The skill I had from before was absolutely no help. I still knew what a vise and bobbin were, but it's a whole different concept in tying. I now teach others, but only the very basics, because I feel I'm still an amateur at this. My students have a lot of knowledge and they know how to handle feathers, and that's very important. They're already at the point it took me four years to reach.

I don't fish for salmon avidly. I enjoy fishing, but I would not kill a fish. I'm not a devoted fisherman; I enjoy watching someone else catch a fish just as much as if I did it, which doesn't make me a dyed-in-the-wool fisherman. It doesn't hold the same fascination as tying flies. Flies are my love. When I do fish, classic salmon flies are all I use—I have a lot of rejects. If they don't get the razor blade they go in the fishing box.

I started out tying classic, established patterns, and I still do a lot. However, I've entered competition in Canada where you send in a "creation" fly, and so I do create patterns for that particular purpose. I'll play around with feathers and combinations of colors, then I'll get an idea and I'll try it. Usually it doesn't work, but it will lead me to something else. Eventually I'll come up with something that I feel is good enough to send to Canada. Right now I'm working on a couple for the next competition.

I got the idea for the Connoisseur Collection [12 framed pieces of artwork, each an original painting surrounded by a dozen classic salmon flies by 12 different tiers, as a fund-raiser for the Penobscot Coalition], and started calling people and asking if they would donate their work. It may not sound like much for one tier to tie a dozen classic salmon flies, but it's a big job. The three artists that I asked all immediately said, "Yes"; they're all from Maine. I think four of the tiers are from Maine. You have to remember, this is a very small group. I don't think there are 12 classic tiers in the state. I'm teaching three now, so we might be up to 10.

They're collectible, and limited of course, and it will never be done again, at least by me. I knew that before I was near through. It took up about six months of my life. But if they sell it will be worth it.

I'm very much an environmentalist. I think the restoration is a very important cause. No matter where I've lived, I've always worked for a cause. I'm also interested in the history of salmon flies, and the history of the Penobscot, since I live on the river.

A year or so ago someone asked me to come to one of the Council meetings, and because no one was there representing the Penobscot club and I'm a member, they asked me if I would sit in as a delegate. I did, and when it came time for new officers to be elected, they asked if I would be the treasurer . . . I have a terrible time saying, "no." As I said, I've always been involved in environmental issues no matter where I lived, always helped raise funds or something for any good group, because I felt that it was important. In each place it's been something different.

Most of my friends have no interest at all in this fishing thing. They do not understand how, if we're trying to restore the salmon run, and I work so hard for this, that fishermen can go out and kill fish. And

I have a terrible time explaining it to them. I think there should be no-kill, and I think they should definitely cut out the Presidential salmon. There should be mention of the first fish caught and released, and it should be released in the name of the President, but to kill a fish doesn't make much sense.

I think more people would support us—the Coalition—if we had no-kill. I think more people would be impressed with us and not Bangor Hydro. There's a lot of controversy, but I know what I hear from people who aren't involved with the salmon, and I think most of the fishermen don't hear that. If you want to restore something, you don't destroy it while you're restoring it. I love to eat salmon too, but you can get it any time of the year now, and the cost is reasonable. We eat pen-raised salmon often, but if I caught one I wouldn't keep it.

I hope the law becomes grilse-only in Maine. I'm sure there's going to be resistance, but the more educated people become, the less resistance there will be. I think it has to be presented in a way that shows how important it is for the restoration. Maybe a presentation to the clubs; if the Penobscot club got together, for instance, and made a unanimous decision in support of it, then the club would stand out and maybe the other clubs would follow too. It would be nice if the Maine Council took a stand.

The fishermen should be grateful the salmon are even back. Let's not foul it up. Don't they want their grandchildren to be able to do what they're doing? I think a lot of the older fishermen—the ones who have been doing it since the restoration started—are not keeping many fish. I think they won't object to catch-and-release. I think it's new blood.

It's here on the Penobscot that it's going to make an impact, not on the other rivers, since it is here that the run is relatively large. I'm not sure the fishermen are going to have much say in it anyway. Even if everyone is against it, if the Department of Inland Fisheries wants to do it, they're going to do it. Even [Atlantic Salmon Commission biologist] Ed Baum has said that the income from salmon licenses is not significant, so that's certainly not a loss [if fishermen stop buying licenses]. I don't even think it will bother the tourists. In Canada it's catch-and-release on many rivers, and there's an awful lot of fishermen on those rivers. Over the long haul, the fishing in Maine should become better than the fishing in Canada. That should be an incentive in itself.

Opening Day on the Bangor Salmon Pool
From the collection of Leroy Hutchings

George Fletcher

At 92, George is one of the oldest (and most colorful) active salmon fishermen on the Penobscot, and he still gets on the river regularly. For a time he ran a fly shop in the Rangeley region.

I caught the Presidential salmon in 1987. I was 87 then—I was born in 1900. That fish was fifteen and a quarter pounds. Brian Summers had been out that year two or three times, but I wouldn't go out because it was so darn cold. Finally we went and had breakfast down to the old club, and we come back up and I said, "I'm going out with you." Brian says, "I'll fish off this side of the boat, and you fish off that side." I don't believe I'd made three casts when *whammo*!

Brian says, "What do you want to do?" I said, "Let's put him back." We started for the pipeline, and Brain and I had seen two near-tragedies there with people going down over the pipeline with fish. Why they ended up alive I don't know, but they did. I says to the fish, "You fella, if you go down over there, you're gonna take a fly with you." I put the wood right to him. By gorry, when I did he came out of the water. "Well," I says, "I don't think we're going to want to put that fella back." You know I tell Claudia, my wife, that I caught a fish and put it back, but unless she can see the fish she doesn't believe me. I said, "We'll take that home to Claudia."

Brian got me on shore, and I supposed he was going to go for the net. I looked around and said, "Where's your net, Brian?" He said, "I'm gonna hand-tail him."

I said, "You stay right where you are, Brian, and I'll let the fish down to you." He did, and he did as good a job of tailing that fish as anyone. But we weren't long in getting him to shore. As long as you don't let the fish hit the water, he's pretty safe, but once he touches that water, boy oh boy, you just as well give up. They're strong fish.

We got ashore, and I said, "By gorry, Claudia will believe this now." Someone come along and said, "That's the Presidential fish, the first fish."

I says, "Claudia's a good Republican, probably she'll be willing to part with it." Claudia wouldn't go down—never get her on a plane anyway—so Brian went with me.

We always went to the Narraguagus, but this is nearer home so we started here about 1980. I knew Fred Thomas. He and Payne and Edwards—those three fellas were probably the three greatest bamboo rod makers, and they were all in Bangor. Just think of it. Payne had this idea about browntoning—he discovered that with a blowtorch he could heat the cane and brown it. Thomas had the idea of putting the guts into a rod. Edwards was just an all-around rod maker. Finally Thomas bought them out. Anybody who has a rod with Fred Thomas's name on it, I'm gonna tell you is some fortunate. And Edwards had two boys, Gene and Bill, that were really wonderful rod makers. Bill Edwards conceived the idea that instead of having six strips in the rod, he'd have four. I had a chance to try those rods out, and I don't know why they didn't go over. People just wanted that six-sided rod.

Going back to this river—I imagine that 200 years ago this was one of the greatest salmon rivers anywhere. But . . . progress. Dams, pollution.

Catching fish isn't the main thing. It's the association with the people. Earl and I go down on the Narraguagus, we always put in a week down there, and if we get one apiece we figure we've been well rewarded and we've done about as well as anyone.

We fish the George River way up in Quebec near Ungava Bay. Now we don't kill many fish. You're allowed to take 10, and we don't take more than four. But just think of it—those engineers know just where they're going to put the dam, and of course that river will be gone. Progress . . . ? They're thinking of putting another dam in here, Basin Mills, and I just hope they don't.

I had a sport shop in Rangeley for about 30 years, and we thought the fishing there was really something. After my son Earl graduated from college, in '57, he said to me, "You know, Dad, I heard about a little river down in Washington County by the name of Narraguagus. They have what they call Atlantic salmon down there. My friend and I will go down there and see what it's like."

I said, "If you go down look up a fella by the name of Harry Smith, because I send him a lot of fly-tying material, and he'll really know what to do."

Well, Harry took him right under his arm. Earl hooked a salmon in the old Stillwater Pool before they put the dam in down there. It was a real good one. But being a trout fisherman, Earl didn't put the wood to him, and the fish got across the river. I'd like to tell you he landed it. He would have if he knew as much about salmon then as he does now. But the fish broke off. He came back and said to me, "You never fly-fished until you've had ahold of one of those." By gorry, he was right.

I went down on the Narraguagus until I got sick of going there, but I couldn't seem to catch a fish. The fellas would call, "Better get down here, George, the river's full of fish." I'd wait a couple of days, and by the time I got down there the run was over. I don't know how many times I did that. Finally I got a chance to go with some fellas from Boston to Portland Creek in Newfoundland. Lee Wulff started things there. They use what they call "the riffling hitch" there. I had some wonderful fishing. Then I came back to the Narraguagus and, by gorry, I used that riffling hitch and I began catching fish down there.

Have you ever seen one of my Frigor Freezes? Down on the Narraguagus, or up on the Matapedia early, it's a great fly. We went up on the Matapedia with Dick Ruhlin and three or four other fellas from here, and they'd been having good luck. They asked us what we had for leaders. We said, "Oh, eight-pound-test, 10-pound-test." They said, "You don't want to use anything but 12-pound-test." I dug around and found some 10-pound.

"What are the best pools?" They named off a few pools. The next morning Earl and I get up and they've gone. We went to all these pools, and never found them on a one of them. We fished all day long, and finally at night we saw them take a fish.

We went down to Adams Pool the next day and tried all the flies they told us about. Well, on the Narraguagus someone had told me that the Mickey Finn is an awful good fly. While I was making some up I made two or three with a pink body. We had tried everything, so I put one of them on.

I got a fish right away, and landed him on the 10-pound-test. It weighed 16 pounds, and by gorry we felt pretty good. Then I got another, about 10 or 12 pounds. We went back and had breakfast, and someone told us where we could take the fish to have it frozen. Earl says, "You know where we saw them catch that fish last night? I'm going down there." I took the net and the camera and so forth, and he went down. I got about halfway down there and I saw that old rod bent up. I threw my rod up in the bushes, took the camera and the net and got down there.

He always said to me, "What I better do Dad, what I better do?" I said, "Put the wood to him." Well he did, and we could see it was a pretty good fish. He finally took Earl out of the pool, down into Salmon Lake. I followed him down with the net. He had to play him all over again after he got down there. When we finally landed that one, it weighed 20 pounds.

We caught several more fish, and took them down to where you could get fish frozen. We were trying to think of a name for the fly. Over the top of the door, it says, "Frigor." And below that "Freeze." I says, "There's a name for the fly. Frigor Freeze." Now I can't do a thing on this pool [the Wringer], but down to the Narraguagus, or on the Matapedia . . . .

It's just unbelievable how these fellas have brought this river back. The Narraguagus was having a hard time way back, and they took some Penobscot salmon and brought the Narraguagus back. When this one got bad and they wanted to bring it back, they took some Narraguagus fish and brought this back. I believe that some of those had Penobscot River genes. Don't you?

This is the cheapest Atlantic salmon fishing in the world. Now don't you believe they could bring back the Kennebec, and possibly the Androscoggin?

Jan Blake

Jan runs a promotional advertising business from her home in Bangor. Full of humor and energy, she's always around when unusual things are happening on the Penobscot—and is sometimes the cause of them, as you'll see.

I'm involved with the Penobscot Coalition—I'm involved with just about everything controversial that has to do with the Atlantic salmon. I want to see it like I've seen it in the past. Every time I went to Augusta when we were fighting Swift River [on the proposal to rebuild Bangor Dam], they managed to get my face on television. It hurt my business a little bit. I had customers who were on the opposition.

I started fishing the Penobscot the very year the salmon came back and the Penobscot Club reopened. About 1974. I fished salmon before then, I think the first time was 1960 or '61, on the Narraguagus River. I caught whatever the terminal illness is, and couldn't get rid of it. I didn't fish but a couple of times last year, or more than two or three times the year before that. But prior to that, I had my share of hooking, if not landing and releasing, salmon.

I once came down over the bank to go fishing down by the Penobscot Club, and Arthur [Taylor] was standing out there in the water. There was a fish rolling every once in a while, and Arthur came out of the water and asked me to take a crack at it. I only made two or three casts and hooked a salmon, and it took off downstream. Then it went around a rock that salmon were famous for wrapping lines around—which I didn't know at the time. I didn't know that you had to keep your rod up there. Joe Floyd came along and said, "You know, you've got to wade out and get your line over that rock."

I started going out and going out, holding the rod as high as I could, and the next thing I knew I was at the top of my waders and started slipping. If I had fallen in, I would have ended up in Winterport. I felt hands go across my waist, and I heard someone say, "You're all right Jan, I've got you," and I thought it was God. I'm not kidding, because I never saw a soul behind me. I thought, *This is it, I'm a believer*. I turned around and it was John Bailey.

He held onto me and we kept going out, and the water went over the top of my waders. We kept going, and I heard him say, "Whoops!" and the water went over *his* waders. At that point Joe Floyd came along in his canoe, and I passed the rod to him so he could try to free up the line. But the salmon was gone. John and I went back to shore and emptied the water from our waders as a friend of mine was coming over the bank. Betty is just the funniest person in the whole world; she came down over the bank and said, "Blake, I'm sorry for all the rotten things I ever said about you."

One of the funniest things happened on a night I had a date. I said to my date, Jerry, "Call me just before you come over." I was fishing up to Beach Pool, which happens to be one of my favorite pools, and Paul Higgins came in with his canoe and asked me if I wanted to go out in the canoe. He could see that there were salmon all over the place, but there was no way I could reach them from where I was.

I got in, waders and all because it was getting late, and the pool was full of canoes, because that was where the salmon were showing. The conversation going from canoe to canoe went something like, "You gonna come up tomorrow night?" "No, I can't tomorrow night. Jeez, I got to take the wife out tomorrow night for dinner. Maybe I'll get out this weekend." And another would say, "Well, I can't come up on Saturday, 'cause if I don't mow the lawn, I'll *never* get to come salmon fishing. My wife will have a fit." Another one said, "Well, Jesus, can't come up Sunday 'cause if I don't stay home with the family . . . . But I'll be up every other night." This was the gist of the conversation that went on as everyone cast toward the salmon.

It started to get a little late, and cold, and finally Paul said to me, "Jan, any time you're ready to go in, let me know." I said, "Oh, yeah," and looked at my watch. I said, "Oh, my God. Paul, I had a date an hour ago!" Paul took me to shore as quickly as he could. I got to the clubhouse, and everyone there already knew about it. I didn't get over that one for two years.

I broke down my rod, threw it in the trunk, bailed into my car, waders and all, and drove to Jerry's. I bailed out of the car, punched the doorbell. When Jerry came to the door I said, "I'm sorry, have you been trying to get me?" He said, "Yes, I've been trying to reach you for an hour." He looked me up and down and said, "Well, I can tell where you've been." I said, "Give me 15 minutes and I'll be ready."

He came over and picked me up and took me someplace to eat. While we were sitting there, all of a sudden the thing hit me and I started to laugh. I told Jerry the conversations that were going on, and he sat there and said, "Well, I can sympathize with the wives."

I think everyone should do their part. If I am going to fish for salmon, then I should do my part to protect them. It just brings out the fight in me to think that people want to mess up this river. I call it a miracle river—where else in the world can you find a salmon river this beautiful flowing between two cities? I get so I want to get my boxing gloves on. I try to convince people, and I've made a lot of enemies doing it.

The thing that bothers me the most is this potential no-kill. I think the people who are pushing no-kill are the fishermen who have fished in a lot of places and caught a lot of salmon, and I don't doubt that they're maybe a little bit sick of it. It's not a big deal to them anymore.

If we had tried to have a no-kill law 20 years ago, you would have seen some fighting from the fishermen who want it now. I think it's a very selfish thing they're doing. I personally feel a two-fish limit makes more sense. Even the one-salmon limit is a little ridiculous. Anglers who are new to salmon fishing should have the opportunity to keep two.

I feel good about the work that's been done here so far, although I don't know what's going to happen in the future, given the economy. But look what's happened with the restoration program—just an explosion. I'm very fearful about the future. I hope they can keep it up. I know they will if the funds are there. The people involved in the restoration are dedicated people—I think they're wonderful.

Late Afternoon at the Wringer
From the collection of Key Bank, Augusta, Maine

Bob Colson

Bob Colson is a professor of accountancy at the University of Michigan at Ann Arbor, and is an ardent fly fisherman for brown trout and steelhead. On his time off, he can usually be found somewhere on the banks of his beloved Pere Marquette River in Michigan.

The name of the river itself holds mystery and fascination for me, evoking the essence of the half-sport, half-religion of angling for Atlantic salmon. That I have seen in my lifetime the return of Atlantic salmon sport-fishing on the Penobscot strikes me as nothing short of miraculous.

I travel from northeastern Ohio to fish the Penobscot. The first time I made the trip my good friend, Earl Fleegler, accompanied me. We were to meet Arthur Taylor at a sport shop in Bangor; arriving early, we drove the Veazie side of the river from behind the hospital up to the dam at Veazie. We could see a solitary figure fishing by the old Bangor Dam. It seemed that he was standing right in the flow of white water spilling over the breach in the old dam. In reality the angler was standing in the Bangor Salmon Pool. We speculated that it was Arthur, but it turned out to be another angler entirely.

Arthur introduced us to many members of the friendly angling community on the Penobscot. We fished with Arthur and Walt Dickson from their boats that afternoon. One of them put me ashore at the Wringer Pool and I immediately discovered the abruptness of the granite ledges on that side of the river. The water is also cold in late May!

The Wringer Pool holds a very dear spot in my heart, both because it represents so appropriately the rebirth of the Penobscot and because it has been so good to me. Named for the wringer washer that once was visible at its upper end, the walk down the creekbed to the pool is studded with the legacy of the dump that used to occupy that bank. Vegetation has grown around the old car bodies and other nonbiodegradable objects up and down that steep, now verdant bank. Rivers can be managed intelligently if there's adequate commitment from people intent on doing so.

Earl and I visited the Wringer early one morning on that first trip to the Penobscot. We threaded our way down the steep creekbed long before first light. Earl set up a two-handed spey rod that generated quite a bit of interest later in the day. He went through the pool first, with a large Mickey Finn. His whoop of surprise a short time after we started filled the chill air and a salmon swirled a few yards from me. The exultation turned in a single breath to dismay—a phenomenon familiar to most Atlantic salmon anglers—when the hook didn't quite catch, even though Earl had felt him good and he had moved quite a bit of water. The sudden droop of Earl's shoulders, his rod tip dropping to the water and his wry smile were—and are—the epitome of the angler's disappointment.

We didn't have the pool to ourselves, however, as other anglers began to join us. There were people from Canada, other northeastern states, along with a fair number of Maine anglers who came to fish for *Salmo salar*. The anglers organized themselves to rotate through the pool, which could handle about six of us at a time. There may have been as many as 12 of us waiting to fish, but the wait was never more than about 45 minutes. It took about 30 minutes to rotate through the pool, covering the holding lies in classic fashion.

That morning was also our introduction to Salmon Billy, one of the river's many characters. Not only did he have a banner fishing day, landing two Atlantics, he did us all a service by keeping the rotation moving at a reasonable clip. Each time an angler in the rotation did not move fast enough to suit him, Billy would blow very loud on a whistle. That whistle blow would always seem to speed things up a bit. I was delighted when he caught the first salmon of the year several years later and presented it in the time-honored fashion to the President.

I have fun fishing the Penobscot. For some reason, I don't feel the same pressure to catch fish when I feel the pull of its water on my legs that I do elsewhere. Maybe the river brings out a certain playfulness in me that is absent in other places. One day I was chatting with an acquaintance at Dickson Pool. He was fishing from his boat; I was relaxing, hunting the sign of fish from the shore. My acquaintance was duly proud of his new vehicle, parked just above me, and continuously sang its praises. As I walked back to the Eddington Salmon Club I had an almost irrepressible urge to pour a bit of oil under the crankcase of that beautiful new sport vehicle. Fortunately, less playful minds prevailed, but that urge—so out of character for me—has given me many minutes of personal intrigue.

Most longtime anglers have experiences that seem to them semi-mystical in nature. I'm certainly in that group who have entered into a series of events that seems so foreign to ordinary experience that you feel you've entered through a portal into a new reality. One of them occurred on the Penobscot in a 24-hour period where time stood still for me.

It started innocently enough. I left my home in Shaker Heights about 10 PM on an early June evening for the overnight drive to Bangor. Arriving at 1 PM the next day, I took a short nap at the Veazie Elms before meeting Arthur and Ruth Taylor for the Eddington Salmon Club banquet. We had plans to go to Quebec the next day, but decided to spend a day on the Penobscot when I won a gift certificate for a license and some equipment at the dinner. (The only place I've ever won raffle items is at Penobscot-related banquets.) We were the only anglers the next morning. Fishing had been slow, with few Atlantics around. After a lobster-roll lunch we drove down to the Bangor Salmon Club and watched the tide come in. With it we could see the silver flash of fish porpoising toward the breach in the Bangor Dam. Not knowing whether they were salmon or some other species, we decided to try our luck again that afternoon.

For us it was a lark. We took only one rod with us, Arthur's Nova Scotia Special, a cane rod he had made several years earlier and that needed a baptism. We had decided that the salmon were waiting for someone fishing cane. Our plan was to take turns trying out the casting qualities of his rod. One of us put on a large Bomber, white deerhair hackled with bright orange; the rod handled the fly beautifully.

And wouldn't you know, a sea-lice-fresh, relatively large hen salmon decided she just had to have that fly while I happened to have the rod in my hand. The adventure of landing that beautiful creature, along with the rest of that story, is recorded in "Nova Scotia Special," published by the *Atlantic Salmon Journal.*

The river, the fish and the people of the Penobscot have been inspirations for me. Not only in the obvious sense that beauty, wildness and human concern "inspire" someone, but in a far more literal and artistic sense. I've published three major stories centered around the Penobscot, its people and its fish. The excitement they create for me is something I wish to share.

Finally, I've learned something important from my associations with people who are concerned about the future of the river. In order for those of us who use a river for sport-fishing to be convincing advocates for appropriate management policy and action we have to demonstrate that we, too, care enough about the resource to regulate our own activities in such a way as to ensure its preservation.

Ken MacLeod

Ken has had a long and varied political career. He spent 10 years in the Maine state legislature, two in the House and eight in the state Senate; his bid for the US Congress in 1964, as the Republican candidate, was unsuccessful. Since 1974 he has been lobbying at the state level on behalf of various clients, including Friends of the Penobscot River.

When I was a kid, my father was only working a day or two [a week] at the mill down at South Brewer, but he was helping a guy who had a couple of salmon weirs. On the third of July, 1933 or '34, I remember netting 79 salmon from one weir on one tide. We just didn't know what to do with all the fish.

My father got a fish for the Fourth, and he was getting a dollar a trip from Petrie, who owned the weir. They sold the rest to local fish markets, A&P and so forth. Petrie got, I guess, eight or 10 cents a pound.

At the same time my father had two smelt nets that he paid rent on to a Swede down in South Brewer, who worked in the mill. We'd go up to the Hughes brothers' gravel pit—they had a couple of piers—and set the nets off there. Yellow sulphur stuff was coming out of the mill, across the river, and you could see this yellow crap coming down the river. We were getting 60, 70, 80 pounds of smelts on a tide. I said, "Pa, how the hell can those fish live with that crap in the river?" He said, "They won't forever." That was in the mid-'30s, and of course they *didn't* last.

I went door-to-door selling the smelts after we got them picked; that's the coldest work in the world next to shucking scallops. I watched a bunch of women shucking scallops down to Buck Harbor five or six years ago on a cold morning. In April, standing on shore with a 100-foot gillnet and picking smelts out—that's cold.

The river was starting to smell bad then. In the late '20s, they were still taking ice out of the river. There'd be skating parties and impromptu hockey games on the ice. But they were putting a lot of crap in. I think Eastern was one of the biggest polluters back then. Great Northern was still there though, in Millinocket and East Millinocket. They must have been dumping a lot of crap in. It wasn't until the '30s and '40s that it really started to show. In the '50s it was pretty dead.

I started fishing for salmon in '73. We weren't economically, socially and culturally deprived in South Brewer when I was growing up, we were *poor*. Which generally didn't lead to salmon fishing. I did a little fishing with nets for alewives coming up the Sedgehunkedunk Stream, and a little bit of trout fishing.

When I first went to Augusta back in the '60s, I was very conservative fiscally, but I always supported the bond issues for sewage treatment plants. You know, they had 90/10 matching back then from the federal government. The federal government was coming up with the money, which they aren't any more. We passed a $25 million bond issue, I think in 1963, for sewage treatment. I always supported that stuff, and then when I got involved in salmon fishing it was a different kind of a fight.

I got involved in the restoration back when I was in the legislature. We passed some pretty landmark legislation for the environment. I got into the legislature in '63, and ran for Congress in '64. That happened to be Goldwater's year, and that wasn't a good year to run. Went back to the legislature in '67 and stayed until '74. I was president of the state Senate in the '70s. They kicked me out of politics and now I make an honest living—I'm a lobbyist.

I've always been very conservative, both in the legislature and now, when most of my lobbying clients are business groups and companies. There were some businessmen in Bangor who wanted to know what the hell I was doing supporting anti-dam legislation, for example [in the fight against the Swift River proposal to rebuild Bangor Dam]. Fortunately, there are also a lot of businessmen who like to fish for salmon. It's a strange dichotomy there when you help the guys who are very conservative in everything except salmon fishing. The argument is made that hydropower is the best type of power environmentally, except for the fish. But we don't need more power right now—we have a surplus.

The river, as it's come back, has been more than just for salmon. I like to sit down at the Penobscot Club and watch the tide either going or coming. Just to see the different groups that are all using this. Kids with spinning rods trying to get a striper or a smallmouth. Young families and older people hiking along the Indian Trail. People in the area who are just enjoying the river.

This year a lot of sand was washed below the ledges down by the Penobscot Club. Christ, there's a beautiful beach there now. I saw people swimming there several times this summer.

On the whole I look on the restoration optimistically. There are people using the river who never used it before, and appreciate what the river is as a clean river. They're not fishermen, and I think we have to build that support. For a while, when the breach in the Bangor Dam wasn't as wide, I guess it was pretty good for kayaking. I mean, I saw these guys sitting sideways; I don't know how they do it, but there were six or seven or eight there every time there was an outgoing tide.

I think a change is taking place for upstream fish passage, but I'm not sure there's enough being done on downriver passage. That, to me, is what the story is. I just don't know how the hell they make it when those smolts go out.

I think the utilities have become more and more aware. They were forced to because of what happened legislatively—there was strong support against their rebuilding that Bangor Dam. I think they're going to be a lot more cooperative than they have been in past years. Bangor Hydro will never build Basin Mills. Bob Briggs, who's the top guy there and a very, very bright buy, married a daughter of friends of ours, and I was talking with him just a couple of weeks ago. I said, "As a rate payer, I've seen my bill go up. When are you going to start saving some money on that Basin Mills thing? Stop pouring bad money after good. In your heart of hearts you know they're never going to allow that to be built." They've got the DEP and the PUC both against them. I was at the hearing over in Bangor when the PUC came up, and it was a horror show. They turned them down flat, and not only that they put a fine on them. But they're still pursuing it. He's a very bright guy, but I think he thinks they're going to prevail.

Bangor Salmon Pool
From the collection of Bill Ballan

Dick Ruhlin

Born in Bangor and educated at the universities of Maine and Wisconsin, Dick is president and owner of a dental supply/practice consulting firm. Founder and first president of Friends of the River (organized to fight Swift River's proposal to rebuild Bangor Dam), he has also been president of the Brewer Chamber of Commerce, the chairman of the Pine Tree Chapter of the Red Cross and state director of the National Ski Patrol. After three terms on the Brewer city council and two as that city's mayor, he is now in his fourth term in the Maine House of Representatives.

My uncle was an active fishermen on the Penobscot in the '30s and very early '40s. I was hanging around the river when I was five or six years old. I guess I was a river rat. I always loved to fish, I always thought Atlantic salmon fishing was the highest form of fishing, even as a kid. Some of the earliest memories I have are from when I was seven or eight years old, watching the old-timers fish for Atlantic salmon, and I can remember I was very impressed with it.

Back in the '40s the radio stations used to announce the first Atlantic salmon to be caught. That was always a matter of great local interest. I think I can remember—I couldn't have been more than 10 or 12 years old—carrying my father's fly rod, wearing a little pair of hip waders and going out there and trying my best at what we now call the back pond. They used to fish for 'em from a boat, and if you did any wading it was on the back pond. In those days it had more of a current to it and a parking area overlooking that was an excellent place to watch salmon jump and watch the fishing. That was just upstream from the Penobscot clubhouse. So I had an early attachment to the salmon that has continued.

I was lucky enough to fish clean rivers when I was very young. I went up to the Miramichi for the first time when I was a freshman or sophomore in high school, and saw what a really clean river was like, a river that had a lot of fish. The thing kind of got in your blood that way. As I grew older I saw the Penobscot become dirtier and dirtier, and the fishing stopped and the smell from the river got worse. I went away to school my last two years, in Wisconsin. When I came back—maybe I picked up some radicalism in Wisconsin—one of the first things I did was look at that river and kind of make a promise to myself that, "By Jesus, my kids aren't going to go through the destruction of this river that I went through," and I was going to set it right. I set that as a goal that I was going to do my best to achieve.

A brown scum formed on the river from one shore to the other. It looked like you could walk across it. There was a stench to it, and it showed if you looked at real estate. All the real estate of lesser value, that people walked away from, was along the riverbanks. All the rundown buildings were on the riverbanks, while all the better homes were up on a hill looking *away* from the river; never toward the river. Now it's just the opposite. Everyone was turning their back to the river and using it as an industrial carriage or sewer. It was just like taking all the sewers in a town and running them into one stream and letting it run right through the middle of town. You can imagine what that would smell like, with feces and everything else being flushed down it, the industrial pollution and what we used to call Harlem River whitefish—condoms floating in the river.

There were a lot of people involved then, some who no longer have any connection with the river. Joe Floyd's one of them, but there were a bunch of others. John Frawley, Bob Baldacci stand out, John Geaghan, who's now passed away. They were just local people, who I knew from one source or another, who had an interest and really felt badly about that river. They looked at it more as a potential positive resource rather than the detriment that it was. We formed an impromptu group called the

Penobscot Pollution Abatement Group in about 1961. With John Frawley's help—he's an engineer—we got to know a lot about the mechanics of pollution: what caused it, what biochemical oxygen demand was, the terminology, and we actually did some studies on it ourselves just to educate ourselves. We went down and fought in the legislature for the Penobscot Pollution Abatement Act of 1965 [the law is now called the Rivers Class Act of 1965], under which the river would be classified. In the old days it was the Water Improvement Commission. They had A, B, C and D for classifications. The Penobscot was classified as a D ("D" signified use as industrial discharge only), and we got that up to Special C, and then B through this bill, and later updates.

We had a lot of help, and a lot of help that would surprise you. We had Great Northern, who could see the handwriting on the wall and came in at that very same time with pollution abatement programs of their own. Muskie was very active on the national scene, and there was fallout on the state level from that. The act passed in 1965, and everyone had to institute plans for pollution abatement programs on the Penobscot River. That was really the first step in cleaning up the river and getting the salmon restoration going.

Shortly after that there was the national bill [the Clean Water Act of 1968]. We had had discussions with Muskie that set up the Penobscot River as a river that would be used as a model nationally as a restoration river where fishways would be important, that type of thing. That was another push.

We didn't think of it that way at the time, but as you look back now you can see the 1968 Model Rivers Act was really an anadromous rivers bill. The pollution levels included all rivers of all types. But when you get to the fishway portions of it, it focuses on anadromous fishes. I don't recall it mentioning salmon specifically.

We fought hard, and got exceptional resistance. We were a little more active then—didn't have large families, didn't have our own businesses started, and had some time. We would be on the radio at least once or twice a week on all the local stations. We wrote editorials, and there were usually a couple or three newspapers in the area we could count on. We'd work with their reporters to make sure they'd ask the right questions. I found them very supportive. The trouble was a lot of people didn't understand the issue. They knew there was a problem, but they didn't understand what created it, didn't understand the terminology. So I would go to the reporters and say, "Look, here's what biochemical oxygen demand means. And here's how it's created, and here's how it's resolved." We got some beautiful newspaper coverage out of it. Then we got a public push. I think the support was latent, it was there, we just kind of focused it. Everybody wanted to do something, but they didn't know what to do.

That's how we got to that stage, and in the meantime I guess I got to like politics a little bit. History and government were my majors in college, so I had an inclination toward it. I got involved in local government, and one of the first things I did when I got on the Brewer city council was to make sure Brewer was first in compliance on the river.

We had our own pollution abatement program in Brewer. We were the first in the nation to take an existing major industrial polluter, the Eastern Corporation paper company, and incorporate their industrial pollutants in with the municipal pollutants and cotreat them. At the time it was very innovative.

At that time treatment was limited to what are called settleable solids; it would get fiber material out and so forth. It still didn't address the caustics. Where Brewer and the river gained was we took a major polluter down directly to a secondary reduction, which is basically a 90- to 95-percent pollution-load reduction to the river—of all causes, not just settleable solids. That also put Brewer way ahead of other communities. As soon as we did that, the next year, salmon were back in the river.

We realized that that would happen. They had just started some stocking, but there had been native fish coming up the river as far as Verona Island. There was an absolute zero-oxygen block around that island in August, or other periods of low water. Some fish would come up before that and then would drop back down. Some fish would come later. By reducing the pollution, we got the oxygen levels to about three parts per million, which is about the minimum. And the Eastern

Corporation closed down temporarily in '68 because of a changeover in corporate management that caused them to stop the pulping operations. The pulping operations, as I recall, produced about 60 percent of their pollution. Sure enough, go back to the Bangor Salmon Pool and there were salmon out there.

Those were the same fish that had been using Cove Brook down in Winterport all along. Cove Brook and Marsh Stream and others that had maintained small natural runs. A lot of people didn't and still don't realize that there were residual fish in those places. I think the Ducktrap River contributed to that also. No fish had been seen in Bangor for a number of years, from about 1955 until '68.

When we got to that point I enjoyed a period of peace. The clubs started coming back. Then when the Bangor Dam proposal came up, someone said, "Well, Dick Ruhlin's had some experience fighting for the river." They asked me to head it up. Those things take 20, 30, 40 hours a week, and I had a larger family, was a partner in a business at that point, just had a lot of commitments. I guess Joe Floyd was instrumental in talking me into doing it. I kind of looked at it with dread. "Jeez, am I crazy?" I knew what we were getting into. I made the prediction at the time it would be five years and would cost $80,000 to $100,000. Unfortunately, my predictions were pretty close.

But once we got into it, frankly, after about a year or two I got so I loved it, loved being back in the saddle. When I finally had to step out at the end because of legislative commitments and other things, I really was kind of sad to leave. I really got so I enjoyed it, enjoyed the fight and enjoyed the company. We had a super group, and I had a lot of pride in it.

We once had a hearing in Augusta, and Ken MacLeod was our lobbyist. He said, "Dick, one of the things you've got to do is make sure you're going to have people to fill the room up." I said, "How big is the room?" He said, "It'll take about 25." I said, "It's not big enough. Go down and tell the chairman we need a bigger room."

Ken said, "Gee Dick, make sure you get the people there for this, because it would look bad if they get a bigger room and you only get it half filled." I recognized that as a possibility for making an impact, so I waited for them to come back and say, "We've got a room for 50." I said, "What's the chairman's name?" It was Jim McBride, and I called up Jim and said, "Jim, you've got to get a bigger room." He said, "How big a room do you need?" I said, "I don't know. Do we have anything bigger?" He said, "Well, there's one there that will hold 100 or 150." I said, "Let me get back to you."

So I purposely let two or three days go by and said, "Jim, you've got to get a bigger room." In those days there was a room in the bottom of the thing [the capitol building] that would hold 300. Ken MacLeod said, "Dick, you won't get 300 people down there to fill that room." Well, we had them standing about 400-plus in the corridors. That really made an impact, and that's what counts in politics.

You know, you could call and tell people in the group—we had a network so everyone would call 10 others—that something was happening and, if you said that you were going to have 300, you could usually guarantee 350. Even during the middle of the work week, when they had to take off from their families or their jobs—and in many cases they were people who worked per hour or per piece, so it cost them money. I tell you, if they had given me room for 600, we would have had 600. Just a super, super group to work with. We're talking about dead-enders here, people who were really true believers, who'd go to the wall for it. Really, in a way, it made the job of being their leader a lot easier. I found my job was more to keep them in check and to direct the flow of their energy, than to arouse energy.

I always made it a point to personally go over and review the names of everyone who paid their dues to Friends of the River. People who really couldn't afford it were there all the time, paying their dues. It was almost the first bill they would pay when it came. I wouldn't have wanted to be their opponent.

As far as rebuilding the dam is concerned, that's dead, and was dead when the DEP held the first hearing in 1985. That's frankly why I didn't bother hanging around anymore—they really didn't need my services. But we still have one thing left to do—and when I say we I'm speaking as a member of the Friends—and that is to remove the old dam. We had three goals, and the final goal was the removal of the existing structure. It's no longer an impediment to salmon migration, but removing it would increase the number of salmon lies and the nursery area.

To me, it's shocking the city of Bangor hasn't been called to account more strongly by the Corps of Engineers, because there's a hazard to navigation there. Boards and huge pieces of debris come off and go down the river. Someday there's going to be an accident there.

The Atlantic salmon and the rivers go hand in hand; I can't see how anyone could love the Atlantic salmon and not be in love with the rivers, or love the rivers and not the salmon, since the salmon are synonymous with a good river. I don't think there's any question in my mind that if the Coalition continues on the path they're on right now, they'll succeed against Basin Mills. They'll succeed, they've got the right game plan, but when they get done, there will be something else.

They thought I was crazy when I said it would cost close to $100,000 [to fight the rebuilding of the Bangor Dam]. We ended up spending I think $80,000 during the main part of the fight, and $100,000 overall. They thought I was a big spender. Boy, have I got news for them. Wait until they see what Basin Mills is going to cost before they're done.

I made a strategic mistake early on, when we had a chance to raise more money and put it in the bank, which is what Don Shields is doing for the Penobscot Coalition. I didn't do that, and consequently at a critical stage we were scrambling around for money when we didn't have to be. In '83 we passed up a whole fund-raising season when we could have raised another $30,000 or $40,000, but I guess my fiscal conservatism came through and I decided we'd raise the money as we needed it. We should have bankrolled it. I notice Don is not making that mistake, and that's smart. At the meeting before the DEP around 1984, we had to send our lawyer up to Toronto at $125 an hour plus travel fees, as an example. It cost us something like $40,000 in a week for the first DEP hearing, including costs of professional consultants, lawyers, etc., but not including time volunteered by the members of the Friends and their supporters.

The problems are not just on the Penobscot. The Miramichi has got its threats now, and it's always had them—chemical spills, clear-cutting upstream. We've got that problem here. The East Branch of the Penobscot is as fine a piece of salmon river as you'll find. You couldn't dream up a better piece of salmon river. But they're up there clear-cutting, and acid rain is a problem. Spawning rubble is something man can't create. That river is nothing *but* spawning rubble. Yet the land up there is cut right down to the primary tree line. On lots of streams they've cut right to the banks so you've got lots of siltation up there.

I don't think we're going to have a successful restoration program on the Penobscot River until we get some handle on the outmigration problem. I've felt that way now for two or three years—that we've reached a plateau from which we will not proceed until we resolve the outmigration problem. On a good year we're getting maybe 200 or 300 adult fish using the East Branch. Over a decade that would be sufficient if we had good outmigration, because the survival rate there is very high. The Miramichi, the Matapedia, the Restigouche do not have a better nursery area than the East Branch of the Penobscot. But those fish are being decimated on the outmigration. They simply run them through the turbines.

It's true we don't have good upstream passage, and the state and FERC are very remiss in not keeping after the licensees to provide adequate fish passage. The majority of our incoming fish are not making it to where they should be. But those fish that do make it upstream are going to be, through natural selection, of finer stock. They've got the habitat once they get there, but what they produce for us is being ground up on the way out. Even with the best upstream passage you're not going to have better than 50-percent efficiency, but even if you went to 80-percent efficiency it's not going to do you any good because you grind them up on the way out.

What you really need are guards for the turbines, flow controls. Buddy Lane, a Bangor Hydro employee who helped adjust the flows at Veazie Dam, showed me one time. Downstream smolts will just follow the current. Above the Veazie Dam, the current is mostly at the racks guarding the turbines, and they will gather and hold there until there is a flushing action. What he would do is shut down the turbines to let the water build up so the smolt would get flushed sideward and around them. So it can be done—he was a salmon fisherman, he cared and he knew what the operation of that particular plant was. You can put in all the fishways you want, but if you don't control the area where the flushing is, then those fish are going to go through the racks. Just to say, "We're going to build a fishway here," doesn't mean the fish are going to get there.

ONE ON AT THE WRINGER POOL

Record Day at the Bangor Salmon Pool

Bud Leavitt

Bud retired as executive sports editor of the Bangor Daily News, *and still contributes a weekly outdoors column. During his 42-year tenure with the newspaper, he was also outdoor editor; he has covered the salmon restoration on the Penobscot, and lent his hand—and his typewriter—to the effort since its beginning. His "Woods and Waters" television program has been running on PBS for 10 years.*

I always begin this spiel with Horace Bond, and as far as I'm concerned I'll go to my grave believing that we would never have the river we have now had it not been for Horace P. Bond.

I remember before the Atlantic Sea-Run Salmon Commission was authorized by the legislature, prior to that my late wife would look out the window and say, "Oh my God, here comes Mr. Bond." Horace would come in here with the same story. He was so persistent—just absolutely dogged—after that legislature for the organization of the commission. He had a Nash automobile, that was '46, '47, and in he'd come. And hell, he was good for three hours. Same story, this persistence, "We're going to do it, Bud, we're going to do it."

I remember one time about '48 or '49, I had a short fuse, and I said, "Horace, it can never be done. I don't give a damn what you say. I appreciate your enthusiasm and everything, but I don't think you'll ever get the support to bring this river back again to anywhere close to what it was."

In my early starting days—I was on the *Bangor Evening Commercial* at the time—the members of the Penobscot Salmon Club voted with some reluctance to make Bill Geagan and me honorary members. Their list included our names at the bottom, and our names were in italics to distinguish us from the other members. Their breakfasts were colorful things as contrasted to today. Wingate Cram, for example, who was president of the Bangor & Aroostook Railroad at the time, and a huge man, God, he must have weighed 300 pounds, would come in with his chauffeured automobile, always with a couple of fresh bottles of Scotch. He'd put them on the table, and they'd get half drunk. I was a kid and I'd look at these people with awe. It was a tightly knit organization, and Horace would be there, and other members and their guests. Adolph Fischer and all those great names of that era.

When the river went to hell, on Opening Day they would slide down over the ice cakes to fish that pool [the Bangor Pool], and with a little luck they might take a half dozen or eight fish. Well hell, it was page one news. The time that first fish was taken—contrasted to today—they'd take it to Jones's Market here and Heywood Jones would crate it up on ice. They'd notify Congressman Frank Fellows or Senator Ralph O. Brewster, or whoever was the ranking member down there at the time, and they'd get it the hell to Washington for the President.

Anyway, then the river went to hell in the early '40s and Horace was trying to restore some sense in state government, hoping to bring back this fish. He was the first person to be chairman of the Atlantic Salmon Commission. He was *it*. I think the only reason they ever created the Commission was to get rid of Horace, because each and every week he was at the legislature hounding the lawmakers. Finally they approved a commission to get Horace the hell out of the way, I always felt. In my judgment, no single individual in the history of the river was more responsible than he, because it was his persistence that really got things rolling.

I remember one day during one of the following seasons, anglers got 24 fish in one day. I wrote a column saying what a pity Horace didn't live to see this day with so many fish being caught. And I wrote we ought to do something by establishing some kind of a memorial. I started the thing off with five bucks. Well, Jesus, two days later, I had to put a box in the paper, "No more money." I had $460. Somebody walked in and gave me a $50 bill, and I didn't know the person. He wouldn't identify himself. He said, "I'm so happy that this river's come back. I want to be anonymous. Here's 50 bucks toward that fund." I made arrangements with Frank up the road here, who sells gravestones. He engraved the rock [in front of the Penobscot Club, with salmon catches for each year inscribed]. Frank chose the rock and we started that thing as a testimonial to Bond. I'm sorry interest has tapered off. The Penobscot Salmon Club has neglected to keep the yearly rod catch current on the Bond rock.

But I go back to then, and to what it is now. To see the number of people who are able to use the river now, from kayakers to canoeists to anglers. You look up one shore and there are more people fishing—say the Eddington side or the Veazie side or whatever—than was the entire membership of the Penobscot Club in its heyday, when they fished only below the dam. I was on a talk show in Hartford not too long ago, and I said I still hold to the fact that this river is one of the great conservation stories that we have on the North American continent. The restoration and the run of fish that has come back—it's cyclical, there are good years and bad years, but that's the nature of the quarry fishermen seek.

With respect to the plans to "bridge" the river for hydropower and so forth, no way, in my judgment, will this ever even the playing field against the value of having this fish come back. They can say all they want to about new modern fishways and new tunnels. You have only to look at Beechwood on the St. John River in New Brunswick. The Tobique. Talk to the people in Canada—this fish just will not navigate comfortably upstream the minute it runs into obstructions. When you put in a fishway, a few fish will use it of course, but most will not. I see it in the long haul as going right back where we were prior to '47. All those rivers attendant to the Tobique, the Serpentine and all the others provided good fishing . . . the Tobique Club was one of the world-famed clubs, and now it's nothing more than a shadow of itself, with a few abandoned buildings.

I remember when they built the golf course at the Aroostook Country Club. The late Perrin Edmonds of Fort Fairfield called me, and wanted to know about getting Sam Snead for the opening round. I had just got through fishing with Sam in the south. I said, "I'll be very glad to ask him, Perrin, but he commands a lot of money." Perrin said, "Well, what do you mean a lot of money, Bud?"

I said, "I don't have any idea, but I'll be the intermediary and ask Sam if he'll do it if he's got that date open." Well I did, and the price was $2,500. Today Sam, even at 80 years old, won't sign anything for less than $10,000. I called Perrin, and he said, "Let's do it.

Snead went up there, and he never was a fly rod man. He's a baitcaster, for bass. We left Fort Fairfield after he played the exhibition match that day, and went with Ray Knight out on the St. John. "Jeez," I said, "Sam, the Mounties are going to arrest you." It was embarrassing. He was throwing that damn bait-casting rod, and he got two salmon. I think I got one that night. We took three like that, and we only had an hour and a half to fish. We never were even down by the bridge near that great shore where the salmon first come into the St. John. We were out in a boat. It was a cinch to go out there and catch a couple of fish—there were hundreds of fresh salmon—but they all stopped when they built the Beechwood dam.

People try to keep secrets. If someone caught a fish and he was over the hill and gone with it, I've always said, "There ain't no way so-and-so can get away with this." I had an example one night, when I was on the Penobscot. I called my wife and said, "I'm going up to the club, I'll be back a little late for supper." This was around four o'clock. I got up there and Dr. Frank Gilly was out on that rock right above the Penobscot Club Pool.

Frank said, "Bud, come on down here and take this fish. I've had him up twice and he won't take." I said, "No, no, I'm not going to fish, go ahead." I'm sitting up there talking, and Frank left the rock. Before he left, he said, "Go on and try that fish." I said, "All right." Well, I had a pair of suntans and loafers on. I went in the clubhouse and picked up my rod. I made a couple or maybe three casts, and the salmon rose and rolled for it. He never touched the fly. I started cheating a little bit, edging down on the rock. Well I slid into the water, right up to my chest. There were a number of witnesses, and hell, I didn't think anything of it. I got back onto the rock, and the sun dried me out.

In the meantime my wife ran out of milk or butter or bread or something, and went down to Joe Lounsbury's grocery store. Lounsbury wasn't at the club, to my knowledge. Yet Joe's laughing like hell, telling Barbara about Bud stepping into the river.

I got home, and I didn't say anything about it. Barbara said, "Was it hot up there?" I said, "No, not particularly. It was beautiful." She said, "Oh, really? Were you perspiring?" I said, "What are you talking about?" She grabbed my shirt, and it was still a little damp, and she says, "I've never seen you come back like this."

I asked, "What are you talking about?" She said, "Oh, nothing, nothing." She kept trying to bait me because Joe forbid her to say anything.

I said, "Where have you been?" She said, "Oh, I've been down to the store."

"Who were you talking to?"

"Joe was there, and I said hello." She couldn't contain herself, and started to laugh. That goddamn story had gotten from the salmon pool to Hampden before I got home. You think you've got something secretive . . . .

You didn't break into the Penobscot Club easily in the early days. Sometimes a member invited you out, like Walter Crossman or Guy Carroll. I went out with Guy a couple of times, with Adolph Fischer once. You can almost remember all the times you had an opportunity to fish. There was very little shore fishing. Everything was by boat.

There was nothing public about that place. It was a public access, but you were met by a cold stare when you drove into the club grounds, even to watch the fishing. It was their part of the world, and any intrusion was looked upon with very little enthusiasm.

Of course that all changed when the dam was breached and opened up the river all the way to Veazie. Not many people realize within five miles of the city of Bangor you have the Penobscot Conservation Club, which is certainly a $250,000 piece of property, the Penobscot Club, the Eddington Salmon Club, and on the opposite shore you have [the] Veazie [Club]. Four great clubs within five miles of downtown Bangor. I doubt very much if there's another major river in the United States where you have that kind of an arrangement. You're looking at at least 1,000 people who belong to those four clubs. I'll tell people from out of town to come up (I won't say from away, or I'll really sound like a Mainer), and point out that they're welcome to go to the Veazie or Eddington clubs, leave their automobile and go down and fish the shore. Or any of the clubs. Nonmembers don't understand that. They think of a golf club or something. They have a hard time believing that you can do this.

I cannot tell you how many times I've seen it . . . some guy just starting out, ill-equipped, his knots are bad, doesn't know how to tie a fly on, and someone will stop and get the guy straightened out. It was Dennis Mason at Veazie one day last season. A man came, and Christ, he had the damnedest outfit you ever saw. Dennis cut his fly line, got the junk off, put a nail knot on, went all through a double Turle knot. The guy couldn't understand that, and said, "Now what do I owe you? Are you a professional?" Dennis said, "Hell no, I'm a carpenter." He took a lot of time, took half or three quarters of an hour, showed the visitor how to get up to the B Pool. The man couldn't understand that. Well that didn't happen years ago. They had the big Leonards, greenhearts, and you never touched their equipment.

The commercial on releasing the salmon that Ted [Williams] and I did [sponsored by J. J. Nissen Bakeries] won the 50th anniversary conservation clip in the United States. There was a news story, and 800 people at the dinner when they awarded the J. J. Nissen Company recognition for that commercial.

John J. Nissen, the old-time owner, is semi-retired but still watches the purse strings of the operation. Ted and I were talking one day and I said, "We ought to do something in the area of conservation." Ted said, "Yeah, good idea." So I went to Mario Fernachio and Ernie Bowler, the script writers. They didn't think it would amount to anything. So I said, "You don't understand." This went on for two, almost three years. Finally Mario wrote some of the text, and Ernie made up a storyboard, and John Nissen happened to be at the planning session.

"That goddamn thing won't sell any bread!" was John's response. "We're in the bread business, not in the send-the-fish-back-into-the-river business." John wouldn't say yes or no, but he had an opinion. It wouldn't sell bread. He called the commercial "Bud's Birds and the Bees." They started playing it, and got swamped with mail. People would be in New Hampshire or Vermont or the Berkshires in the summer on vacation, and they'd see this commercial, and these people were all concerned with the environment. The company got hundreds of letters, complimenting them on this conservation effort for clean water and an opportunity to fish and all that. Suddenly, John said, "Well goddamn it, it worked. It cost us money and we didn't sell any bread, but it worked." Then we did the other one on litter. Now we're going to do five more. People were writing in and saying, "I've never had a loaf of your bread, but that commercial has sent me to your bread. As long as you have a concern for the environment, I'll always buy J. J. Nissen bread." That was the tone of some of the letters. Well, Jesus, that made old John feel like a million bucks.

This poor fish, as long as we live, as long as our descendants live, will always be a target. Always has been. From poaching to long-lining to commercial fishing.... It has foes from coal-fired arrangements in Bucksport all the way to Iceland, under the ice cap, wherever the hell they go. From the time it's born until you put it away, if it escapes it's a miracle. It's nice that this [the buyout of commercial licenses] has been done in Newfoundland, but it always has foes. Somebody will come up with a goddamn dam proposition, someone will want to bring a coal-fired plant in... anything they do will put spillage in that river.

I feel sad to think that anything might happen to the Penobscot or any other Maine river. I've only got one grandson, and the very thought he might not go to the Penobscot and catch an Atlantic salmon—it bothers me to think that this river could possibly be reverted back to what it was in the middle 1940s if plans go through.

It's a pity that we don't have a blue ribbon group of concerned people who would go to Montana and see what that state has done to its rivers, how they've preserved fishing with stringent rules and regulations. The only way they can do it is because of splendid programs. Look at what they've done in Wyoming, what they've done in Idaho. It's been my good luck to have that experience out there every year for the last 15. They've got management, they've got leadership in state government. If you talk Atlantic salmon in Augusta, hell, they're all pickerel fishermen, it doesn't mean anything. And I've got nothing against pickerel fishing, but they don't recognize that the salmon's pretty special.

Last August when I was out there [in the West], for example. We went to a different place each day, and when we'd stop in a store there'd be eight fishermen, three fishermen, five fishermen. Stop to get gasoline, you'd be the third or fourth car . . . rods on the roof, rods in the back. They're going somewhere—they were going to West Yellowstone, the Gallatin, or possibly the Henry's Fork. They're going fishing, and it costs money. Everybody's aware of it.

There are no two cities in America with a major Atlantic salmon river flowing between the two of them—other than Bangor and Brewer. You can be in your office all day, or you can pump gas or you can sell shoes, or you can work in the Eastern Corporation or wherever, and at four o'clock in the afternoon if you want to go fishing you can throw a rod in your car and go to the Penobscot and fish. You can't do that anywhere else for Atlantic salmon—you know the price of this privilege in Canada.

There's this feeling that this is an elitist fish. Well, good Lord, you see people with equipment valued at nine dollars. And they'll hook a fish and land it. And you see people with thousand-dollar rigs. One rod has just as good a chance as the other. You can come in a Rolls-Royce or in a pickup truck.

I'll take people fishing for a couple of hours and, coming home, they'll say, "Boy, you're lucky. Do you realize how lucky you are?" And I'll say, "I realize it." And they'll ask, "But do the people realize it?"

I answer, "No."

Gordon Brook on the Mattawamkeag

On the River

At 5:30 in the morning, even on a still day that will see temperatures in the 90s, the chill at the Eddington Pool on the Penobscot River is biting. Anglers awaiting their turn in the rotation stuff their hands down the front of their neoprenes in an attempt to keep them warm or, if they've thought to wear a jacket—I'm not one of these—keep their hands in their pockets. I put my rod in the rack, noting there are six or so fishermen (one is a fisherwoman, as a matter of fact) ahead of me. That means a wait of perhaps two hours.

That wait isn't so bad. As six o'clock approaches late arrivals, knowing they have hours to wait, deposit their rods in the rack and head a half mile or so upstream to a public breakfast given by the Eddington Salmon Club. They will come back for their turn when the temperature and, they hope, the fish, are hotter—it's been slow for this late in May. Those who stay talk about this season, seasons past and, of course, the Red Sox. I listen—you never know what you might pick up and, besides, I'm a Mets fan. Better to keep my own counsel.

At about 7:30 my turn comes up and the angler behind, John (didn't catch his last name), accompanies me. He's a member of the Eddington club and apparently did much of the volunteer work to build the road down to the pool. He knows I've never fished through Eddington Pool before, and wants to show me how to fish the precarious upper 50 yards or so—it's a stretch that, apparently, many people don't bother with. Here we're very close to the Veazie Dam, and the strange angle at which it crosses the river's flow makes the current tricky. Three or four rugged rock fingers angle into the current and downstream. They make convenient casting platforms, but at this water level they're under the surface and good places to trip as well.

John does everything but take me by the hand to show me how to get out into the run, then points out where to step to fish the rest of the dangerous stretch and, from there, the remainder of the pool. Sharing is what this river is all about. There are really only about three or so miles of fishing to be had (and a limited number of good pools in those few miles), and we're within spitting distance of two cities with a combined population of more than 40,000. Cooperation among anglers is important, here on the river as well as when arguing the Penobscot's case to state legislators in the State House in Augusta and at hearings before various regulatory agencies.

Cast, step, step. Cast, step, step. The Eddington Pool may be the finest fishing pool for Atlantic salmon in the United States. On the left bank of the river, it's tailor-made for a right-handed caster, and gives up a large percentage of the fish caught annually on the Penobscot. If Bangor Hydro's addition of another power station to the Veazie Dam just a hundred yards or so upstream is approved, this pool will likely become a featureless back eddy.

I work through the pool, maintaining a consistent distance from the angler rotating through in front of me. Nothing happens—except that I manage to lose the General Practitioner I'm casting by ticking it on the rocks behind me (got to keep that backcast up), and change to a Verdict, a popular fly developed by Jerry Clapp for this river. I get to the bottom of the pool at about 9:30, reel up and head back to camp to meet Arthur. Later in the day I learn that, at about 10AM, a fisherman hooked and lost a good fish in the lower Eddington Pool. He couldn't have been more than two or three anglers behind me in the rotation—I had probably cast over that very salmon. Maybe I should have gone to that breakfast.

In spite of some dire forecasts, and a disappointing salmon return in 1991, things are looking up for the Penobscot. The anglers and fishery professionals we've heard from who have the longest-reaching outlooks all believe the Atlantic salmon will be restored to the river, if—and this is a big if—the critical problems of upstream and downstream passage are dealt with. Progress is made all the time. Consider what appears to be the effective end of the Newfoundland and Labrador commercial fishery: In March of 1992 the Canadian government announced a voluntary buy-out of commercial licenses in Newfoundland, including a five-year fishing moratorium. The cash payments were also offered to Labrador commercial fishermen, and the Atlantic Salmon Federation expects many netters to take advantage of the program. ASF President Dave Clark said, "I predict this marks the end of the Newfoundland and Labrador commercial salmon fisheries." A marked reduction in, or elimination of, commercial salmon interception off Newfoundland and Labrador could mean a significant increase in returns to the Penobscot.

The fact that hundreds of dams nation-wide will be up for relicensing over the next several years is a distinct plus for those who wish to protect rivers, too. It's possible that Bangor Hydro will be allowed to add C-Station to the Veazie Dam on the Penobscot, for instance, but it's likely that a stipulation of any relicensing will be significant concessions to upstream and downstream passage for anadromous fish. There is little doubt that building C-Station would radically change the salmon fishery below the Veazie Dam, if only by significantly altering salmon lies. While it's possible that the salmon would simply shoot through the stretch of the Penobscot between the Bangor and Veazie dams, providing little or no recreational fishing, Ed Baum points out that such a change in flow in the river might make conditions better in general. By concentrating the river flow and reducing the amount of spill over the dam at Veazie, salmon might find the fishways more easily and get upriver to spawn more quickly. But if a new power station and Basin Mills are both allowed to go through, the Penobscot fishery might be destroyed entirely.

(Eternal vigilance is the only weapon against the constant threats to the river. Listen to Ed Baum's comments on Bangor Hydro's attempts to get approval for the construction of the Basin Mills project: "Whenever someone wants to build a new dam they have to advertise, to see if someone will sell them the power cheaper. I read an article recently that said all the responses Bangor Hydro had were higher. They said the cheapest was eight cents a kilowatt/hour, and with a new dam they can generate it for six cents a kilowatt/hour. I was at a meeting a couple of weeks ago, and a fellow from the PUC [Public Utilities Commission] on our committee told us Central Maine Power was buying power from out of state at one and a half cents a kilowatt/hour. There's a glut of it." Eternal vigilance, indeed.)

The recent recreational-angling regulation changes in Maine mean fishermen are doing their part, too. With luck, a one-fish limit on Atlantic salmon will mean more wild fish will make it upstream to the spawning grounds—and every wild fish on the spawning beds brings us closer to a truly wild, self-sustaining run of salmon.

Thanks to the tireless work of those we've heard from in these pages, and hundreds of anonymous others, the Penobscot flows cleaner now than it did 45 years ago, when Horace Bond was pestering the state legislature to establish an agency whose mission it was to bring the Atlantic salmon back to Maine. And there are salmon in the river again. Enough to bring anglers back to the Penobscot on May 1, hoping to fill their salmon tags so they can have the traditional meal of Atlantic salmon and fresh peas on July Fourth (even if they release the fish they catch and reward themselves by buying one from a local market). Enough to get a few upriver to the ancient spawning grounds, to help reestablish a wild stock that will again make the Penobscot its home river. And enough so that one will almost certainly find its way to the White House table some time in May of this year—and for many years to come.

—J.E.B.

Bibliography

Magazine, Newspaper Articles

Anderson, Mark W. "Is Hydro Power Boon or Bane?" *Bangor Daily News*, date unknown.

Averill, Joni. "Aileen Shea Zahn Remembers the Pool." *Bangor Daily News*, June 10, 1981.

Baum, E.T. "1991 Maine Atlantic Salmon Program Summary." *The Connecticut River Salmon Association Newsletter*, Winter, 1991.

Bray, Sharon. "Craig Brook National Fish Hatchery." *Castine Patriot.* October 5, 1989.

Chamberlain, Tony. "Comeback Success is Catching." *The Boston Globe*, June 15, 1991.

Chamberlain, Tony. "It's Praised to the Rafters." *The Boston Globe*, June 16, 1991.

Chamberlain, Tony. "Rebirth of a River." *The Boston Globe*, June 14, 1991.

Dietrich, Mary Lou. "He Carried Gun and Flute." *Bangor Daily News*, June 28, 1984.

Hennessey, Tom. "Brewer's 'Bangor Salmon Pool' has Rich Angling Heritage." *Bangor Daily News*, June 29, 1989.

Hennessey, Tom. "Ed Houston." *Bangor Daily News*, August 14-15, 1982.

Hennessey, Tom. "Entire Eddington-Veazie Salmon Fishery Threatened." *Bangor Daily News*, March 11, 1992.

Hennessey, Tom. "Plan to Enhance Salmon Restoration Would Need Funds." *Bangor Daily News*, January 11-12, 1992.

Hennessey, Tom. "Thomas Rod Company." *Bangor Daily News*, August 14-15, 1982.

Hewitt, Rich. "Craig Brook Hatchery to Mark 100th Anniversary." *Bangor Daily News*, October 4, 1989.

Hidlay, William C. "High Court Upholds Decision Rejecting Bangor Hydro-Electric Dam Projects. *Bangor Daily News*, date unknown.

Jordan, Richard M. "Atlantic Salmon Fishing in Maine." *Maine Fish and Wildlife*, Summer, 1977.

Kendall, William C. "History of Salmon Angling in New England." *The Salmons*. Boston Society of Natural History, Volume 9, No. 1, 1935.

Leavitt, Bud. "24-Salmon Haul on the Penobscot the Largest in a Long Time." *Bangor Daily News*, June 1978.

Leavitt, Bud. "Never Like This in the Old Days." *Bangor Daily News*, June 4, 1980.

Leavitt, Bud. "The 'Reel' Fishing Season Begins Thursday." *Bangor Daily News*, April 29, 1986.

Leavitt, Bud. "The Penobscot: It's a Miracle." *Bangor Daily News*, May 26, 1991.

Lovell, John. "Saving Salmon." *The Maine Sportsman*, September, 1991.

Meister, Alfred L. "Atlantic Salmon Fishing: A Look Back." *Maine Fish and Game*, Spring, 1964.

Pennington, Samuel. "Sporting Collecting Grows." *Bangor Daily News*, February 17, 1978.

Radigan, Kyle. "Dam it All." *Trout* Magazine, Summer, 1991.

Shields, Don. "Basin Mills Update." *The Rusty Rat*, Winter, 1991.

Smythe, Olive M. "Historical Account of Fishing Activity at Famous Bangor Salmon Pool." *Bangor Daily News*, April 1, 1953.

Sullivan, Edward C. "The Day a Man's Boyhood Dream Came to Life at the Veazie Dam." *Bangor Daily News*, December 18-19, 1976.

Swartz, Brian. "For More Than a Century, Shipbuilding Reigned Supreme." *Bangor Daily News*, June 29, 1989.

Townsend, Clinton B. "Hydroelectric Industry Thwarts Salmon Restoration." *Bangor Daily News*, August 14, 1991.

Weber, Tom. "The Presidential Salmon Tradition Began in 1912." *Bangor Daily News*, June 28, 1984.

——. "Bangor's Famous Salmon Pool." *Maine Woods*. Bangor & Aroostook Railroad. 1938.

——. "Basin Mills Project Temporarily Stopped." *American Rivers Annual Report*, 1990.

——. "Maine Can Examine Fishway, Public Access Adequacy, Court Rules." *American Rivers*, Fall, 1991.

——. "Penalties Urged Against Bangor Hydro." *Bangor Daily News*, April 16, 1991.

Pamphlets, Papers

Atkinson, Clinton. "A Chronological History of the Bucksport Hatchery." Date unknown.

Baum, E.T. "Maine Wild Atlantic Salmon Protection Policy," (draft). 1990.

Cutting, Richard E. "Penobscot River Salmon Restoration." Maine Atlantic Sea-Run Salmon Commission. 1959.

Everhart, W. Harry and Cutting, Richard C. *Penobscot River Atlantic Salmon Restoration: Key to a Model River*. Washington, D.C.: Dept. of the Interior, US Fish & Wildlife Service. 1967.

Everhart, W. Harry, Watson, John E. and Cutting, Richard C. *Penobscot River Salmon Restoration*. Augusta, Maine: Atlantic Salmon Commission. 1955.

Hauptman, Cliff. *The Atlantic Salmon in Maine: A Complex and Valuable Resource*. Maine Council, Atlantic Salmon Federation. 1989.

Kendall, William C. "Founders of Fish Culture: Charles Grandison Atkins." 1936.

Monaghan, Joanne. *Our Penobscot River Heritage*. Bangor, Maine: Gordon Clapp Travel Services. 1981.

Rounsefell, George A. and Bond, Lyndon H. *Salmon Restoration in Maine.* Augusta, Maine: Atlantic Sea-Run Salmon Commission. 1949.

——. *The Atlantic Salmon in New England.* Ipswich, Massachusetts: The Atlantic Salmon Federation. 1990.

——. *Bangor River Days.* Bangor, Maine: Penobscot Heritage Museum. Date unknown.

——. *Craig Brook National Fish Hatchery.* Washington, D.C.: Dept. of the Interior, US Fish & Wildlife Service. Date unknown.

——. *The Founding of Century-Old Craig Brook National Fish Hatchery.* Washington, D.C.: Dept. of the Interior, US Fish & Wildlife Service. Date unknown.

——. *The Golden Era of Logging on the Penobscot.* Bangor, Maine: Penobscot Heritage Museum of Living History. 1970.

——. "New England Atlantic Salmon Restoration Conference." The International Atlantic Salmon Foundation World Wildlife Fund. 1975.

——. "Newfoundland Commercial Salmon Fishing Closure," (press release). St. Andrews, New Brunswick: The Atlantic Salmon Federation. 1992.

——. *News and Views of Maine's Atlantic Salmon* (newsletter). Orono, Maine: Atlantic Sea-Run Salmon Commission. March, 1960.

——. *A Visitor's Guide: Green Lake National Fish Hatchery.* Washington, D.C.: Dept. of the Interior, US Fish & Wildlife Service. 1985.

Books

Echeverria, John D., Barrow, Pope and Roos-Collins, Richard. *Rivers at Risk: The Concerned Citizen's Guide to Hydropower.* Washington, D.C.: Island Press (American Rivers). 1989.

Keane, Martin J. *Classic Rods and Rodmakers.* Stockbridge, Massachusetts: Classic Publishing Company. 1976.

Index

I-J-K

L

M

N-O

P-Q

R-S

T-U

V

W-X-Y